The Sustainable Composite Materials in Civil and Architectural Engineering

The Sustainable Composite Materials in Civil and Architectural Engineering

Editors

Yeou-Fong Li
Walter Chen
Ta-Wui Cheng

MDPI • Basel • Beijing • Wuhan • Barcelona • Belgrade • Manchester • Tokyo • Cluj • Tianjin

Editors

Yeou-Fong Li
National Taipei University of Technology
Taiwan

Walter Chen
National Taipei University of Technology
Taiwan

Ta-Wui Cheng
National Taipei University of Technology
Taiwan

Editorial Office
MDPI
St. Alban-Anlage 66
4052 Basel, Switzerland

This is a reprint of articles from the Special Issue published online in the open access journal *Sustainability* (ISSN 2071-1050) (available at: https://www.mdpi.com/journal/sustainability/special_issues/materials_engineering).

For citation purposes, cite each article independently as indicated on the article page online and as indicated below:

LastName, A.A.; LastName, B.B.; LastName, C.C. Article Title. *Journal Name* **Year**, *Volume Number*, Page Range.

ISBN 978-3-0365-3501-2 (Hbk)
ISBN 978-3-0365-3502-9 (PDF)

Contents

sustainability

MDPI

Editorial

The Sustainable Composite Materials in Civil and Architectural Engineering

Yeou-Fong Li [1], Walter Chen [1,*] and Ta-Wui Cheng [2]

1 Department of Civil Engineering, National Taipei University of Technology, Taipei 10608, Taiwan; yfli@ntut.edu.tw
2 Institute of Mineral Resources Engineering, National Taipei University of Technology, Taipei 10608, Taiwan; twcheng@ntut.edu.tw
* Correspondence: waltchen@ntut.edu.tw

Citation: Li, Y.-F.; Chen, W.; Cheng, T.-W. The Sustainable Composite Materials in Civil and Architectural Engineering. *Sustainability* **2022**, *14*, 2134. https://doi.org/10.3390/su14042134

Received: 7 February 2022
Accepted: 10 February 2022
Published: 13 February 2022

Publisher's Note: MDPI stays neutral with regard to jurisdictional claims in published maps and institutional affiliations.

Sustainability is a central value in the United Nations' 17 sustainable development goals (SDGs), which include no poverty, zero hunger, good health and well-being, quality education, gender equality, clean water and sanitation, affordable and clean energy, decent work and economic growth, industry innovation and infrastructure, reduced inequalities, sustainable cities and communities, responsible consumption and production, climate action, life below water, and life on land. The use of sustainable composite materials in civil and architectural engineering, which is the theme of this Special Issue, is particularly related to SDG #9 (building resilient infrastructure, promoting inclusive and sustainable industrialization, and fostering innovation) and SDG #11 (making cities and human settlements inclusive, safe, resilient, and sustainable). We believe that, of all feasible solutions, the use of sustainable composite materials is critical to accomplishing these aims.

However, when the *Sustainability* journal approached us to launch this Special Issue, the world was facing an unknown infectious epidemic and quickly fell into a state of uncertainty as a result of the COVID-19 pandemic's rapid spread. Two years into the outbreak, there appears to be no end in sight. We cannot help but wonder what lies ahead. Everything has altered as a result of the unexpected pandemic, from how we keep ourselves safe to how we connect with others. It is difficult for people from all walks of life, not just academics. According to the United Nations, the epidemic has exacerbated poverty and weakened our ability to respond to long-term sustainability challenges. It is a sobering reminder that even in the face of a pandemic, we cannot abandon environmental protection. This is why the Special Issue "Sustainable Composite Materials in Civil and Architectural Engineering" was initiated. To act even in the midst of a disaster, we encourage researchers to continue their work during this trying period and invite them to submit new work on the experiment, analysis, inspection, and repair of a variety of infrastructures and engineering structures to us. Additionally, we purposefully broadened the scope. The composite material does not have to be of a certain type, such as fiber-reinforced plastic (FRP) and geopolymer, but may be composed of recycled and reused waste materials. Furthermore, the contribution may focus on any aspect of sustainable composite materials in civil and architectural engineering, such as carbon emissions, cost analysis, experimental verification, flame retardance, reinforcement, and energy consumption.

As a result, we received 18 submitted manuscripts that covered a broad range of subjects and were excellent pieces of work. However, only ten manuscripts were accepted following a rigorous peer-review procedure. Three of the manuscripts were published in 2020, six in 2021, and the final article was published in 2022. The following is an overview of these works.

Lee et al. [1] conducted a study on oxygen furnace slag, a significant waste by-product of steel manufacturing that can be used as a natural aggregate. To minimize undesirable expansion, the authors applied novel geopolymer technology to capture free CaO and

MgO and build a stable silicate composite. The laboratory- and plant-scale experiments showed that expansion might be controlled to less than 0.5 percent while increasing the compressive strength of the specimens.

Focusing on a different material and application, Cheng et al. [2] detailed the development of conductive films using silver (Ag) flake powder and a multiwall carbon nanotube (MWCNT) hybrid grid on a polytetrafluoroethylene (PTFE) film for applications that require both electromagnetic shielding (EMS) and a conductive film. The suggested conductive film exhibited superior electromagnetic shielding effectiveness (EMSE) and conductivity compared to materials containing varying quantities of MWCNT. Additionally, the film was more stretchable, with a 10% elongation at a 29% resistivity change rate.

In the context of recycling materials, Lin et al. [3] investigated and tested the usage of SiC sludge as a geopolymer material in place of metakaolin. The mass ratio of Na_2SiO_3 and NaOH solutions (NS/SS ratio) was experimented with, as well as the influence of SiC sludge content on metakaolin geopolymers. The results indicated that the geopolymer with the ideal SiC sludge replacement level and NS/SS ratio possessed a high heat evolution value, superior flexural strength, and high Q4 silicate deconvolution percentages, which is due to a synergistic effect, increasing both the reactivity and strength of metakaolin-based geopolymers. As a result, SiC sludge qualifies as a potentially useful ingredient in the production of geopolymers containing metakaolin.

The following article by Shen et al. [4] contributes to the conversation regarding an eco-friendly polyurethane hyperbranched hybrid made through the sol–gel process. The organic–inorganic hybrid material was created by introducing a non-halogenated, hyperbranched flame retardant containing nitrogen, phosphorus, and silicon to a polyurethane (PU) matrix. Using a variety of instruments, the study examined the organic and inorganic dispersity, morphology, and flame retardance mechanism of the hybrid material. The hybrid material not only had a high condensation density, but it also had outstanding organic–inorganic phase compatibility. Finally, the hybrid material passed the burning test and demonstrated outstanding flame-retardant properties.

Shen et al. [5] investigated two dispersion methods (planetary centrifugal mixing and three-roll milling) to minimize the time required to disperse graphene nanoplatelets into a polymer matrix for graphene nanoplatelet-reinforced epoxy nanocomposites. Ultimate tensile strength, flexural strength, and flexural modulus were used to evaluate the procedures. The results revealed that planetary centrifugal mixing is more effective. It not only took less time to complete, but it also produced a more uniform dispersion of graphene nanoplatelets than three-roll milling and other traditional dispersal methods.

Wang et al. [6] investigated the effect of nano-MgO content and carbonization time on nano-MgO-modified cement soil utilizing mechanical characteristics as well. The results revealed that by adding 1.0 percent nano-MgO to the modified cement soil and carbonizing it for one day, the compressive strength of the modified cement soil may be greatly increased. The energy dissipation rate of the modified cement soil after 1-day carbonization achieved its maximum, and the peak strain of the modified cement soil after 2-day carbonization reached its highest value, both with the same nano-MgO concentration.

Returning to the topic of furnace slag, Li et al. [7] investigated the radiation cooling effect of substituting basic oxygen furnace slag (BOFS) on asphalt concrete pavement. Thermal conductivity, emissivity, and indoor and outdoor temperature measurements were used to compare the thermal performances of varied proportions of 45 wt%, 55 wt%, and 75 wt%. The specimen with the BOFS substitution of 75 wt% absorbed the most heat inside the body, resulting in less heat being released into the environment as a result of this. After making the appropriate BOFS substitution, the specimen's stability value, indirect tensile strength, and British pendulum number (BPN) all met the criteria for each parameter. Finally, because of its thermal performance, BOFS offers a wide range of potential benefits in pavements, particularly for the purpose of achieving the goal of urban heat island mitigation by radiation cooling.

Carbon-fiber-reinforced plastic (CFRP) has been widely employed in civil and architectural engineering to repair or replace deteriorated or damaged engineering structures such as bridge decks, concrete beams, walls, slabs, and columns. Its adaptability has sparked a wave of invention and several technological advancements. Li et al. [8] employed microwave-assisted pyrolysis (MAP) to remove the resin from a CFRP bicycle frame in order to recycle waste carbon fiber. They analyzed the mechanical properties of carbon-fiber-reinforced concrete (CFRC) in static and dynamic testing using three distinct types of carbon fiber, which are conventional carbon fiber, carbon fiber without a coupling agent, and recycled carbon fiber. The mechanical performance of recycled carbon fiber was found to be superior to that of normal carbon fiber and practically identical to that of carbon fiber without a coupling agent.

Li et al. [9] studied the influence of a next-generation colored inorganic geopolymer material (IGM) paint on an insulating concrete building shell for another geopolymer use. Five insulating IGM paints, white, red, green, blue, and yellow, were applied to the top surface of a concrete slab to determine their ability to reflect heat and to reduce a building's cooling requirements during hot summer seasons. The results indicated that IGM paints significantly reduced the surface temperature and heat flow of the upper and lower surfaces of concrete slabs, with the white IGM paint had the best performance of the five colors.

Chin et al. [10] closed the Special Issue with another investigation on waste CFRP. The authors investigated the undesirable silane residue on recovered carbon fibers (rCF) that was applied as a coupling agent during the commercial CFRP manufacturing process. The surface morphologies and elements present on the rCF were studied using the microwave pyrolysis method. It was discovered that increasing the pyrolysis temperature results in a greater reduction in silicon content. Due to the uniformity of microwave pyrolysis recycle treatment, this is a viable alternative to traditional furnace technology.

Author Contributions: Conceptualization and writing, Y.-F.L., W.C. and T.-W.C. All authors have read and agreed to the published version of the manuscript.

Funding: This research received no external funding.

Conflicts of Interest: The authors declare no conflict of interest.

References

1. Lee, W.-H.; Cheng, T.-W.; Lin, K.-Y.; Lin, K.-L.; Wu, C.-C.; Tsai, C.-T. Geopolymer technologies for stabilization of basic oxygen furnace slags and sustainable application as construction materials. *Sustainability* **2020**, *12*, 5002. [CrossRef]
2. Cheng, H.-C.; Chen, C.-R.; Hsu, S.-H.; Cheng, K.-B. Electromagnetic shielding effectiveness and conductivity of PTFE/Ag/MWCNT conductive fabrics using the screen printing method. *Sustainability* **2020**, *12*, 5899. [CrossRef]
3. Lin, K.-L.; Lo, K.-W.; Cheng, T.-W.; Lin, W.-T.; Lin, Y.-W. Utilization of silicon carbide sludge as metakaolin-based geopolymer materials. *Sustainability* **2020**, *12*, 7333. [CrossRef]
4. Shen, M.-Y.; Kuan, C.-F.; Kuan, H.-C.; Ke, C.-Y.; Chiang, C.-L. Flame retardance and char analysis of an eco-friendly polyurethane hyperbranched hybrid using the sol-gel method. *Sustainability* **2021**, *13*, 486. [CrossRef]
5. Shen, M.-Y.; Liao, W.-Y.; Wang, T.-Q.; Lai, W.-M. Characteristics and mechanical properties of graphene nanoplatelets-reinforced epoxy nanocomposites: Comparison of different dispersal mechanisms. *Sustainability* **2021**, *13*, 1788. [CrossRef]
6. Wang, W.; Zhou, H.; Li, J.; Tao, F.; Li, C.; Qian, B.; Jiang, P. Influence of carbonization process on the mechanical properties of nano-MgO modified cement soil. *Sustainability* **2021**, *13*, 3558. [CrossRef]
7. Li, Y.-F.; Yang, P.-A.; Wu, C.-H.; Cheng, T.-W.; Huang, C.-H. A study on radiation cooling effect on asphalt concrete pavement using basic oxygen furnace slag to replace partial aggregates. *Sustainability* **2021**, *13*, 3708. [CrossRef]
8. Li, Y.-F.; Li, J.-Y.; Ramanathan, G.K.; Chang, S.-M.; Shen, M.-Y.; Tsai, Y.-K.; Huang, C.-H. An experimental study on mechanical behaviors of carbon fiber and microwave-assisted pyrolysis recycled carbon fiber-reinforced concrete. *Sustainability* **2021**, *13*, 6829. [CrossRef]
9. Li, Y.-F.; Xie, Y.-X.; Syu, J.-Y.; Huang, C.-H.; Tsai, H.-H.; Cheng, T.-W.; Chen, Y.-C.; Lee, W.-H. A study on the influence of the next generation colored inorganic geopolymer material paint on the insulation measurement of concrete building shell. *Sustainability* **2022**, *14*, 164. [CrossRef]
10. Chin, K.-Y.; Shiue, A.; Wu, Y.-J.; Chang, S.-M.; Li, Y.-F.; Shen, M.-Y.; Leggett, G. Studies on Recycling Silane Controllable Recovered Carbon Fiber from Waste CFRP. *Sustainability* **2022**, *14*, 700. [CrossRef]

sustainability

Article

Studies on Recycling Silane Controllable Recovered Carbon Fiber from Waste CFRP

Kai-Yen Chin [1,†], Angus Shiue [1,†], Yi-Jing Wu [1], Shu-Mei Chang [1,*], Yeou-Fong Li [2], Ming-Yuan Shen [3] and Graham Leggett [4]

1 Department of Molecular Science and Engineering, Research and Development Center for Smart Textile Technology, National Taipei University of Technology, Taipei 106, Taiwan; f104gstyle@gmail.com (K.-Y.C.); angusshiue@gmail.com (A.S.); knight32671@gmail.com (Y.-J.W.)
2 Department of Civil Engineering, National Taipei University of Technology, Taipei 106, Taiwan; yfli@mail.ntut.edu.tw
3 Department of Mechanical Engineering, National Chin-Yi University of Technology, Taichung 411, Taiwan; myshen@ncut.edu.tw
4 LI-COR Biosciences, Cambridge CB4 OWS, UK; graham.leggett@licor.co
* Correspondence: f10914@mail.ntut.edu.tw
† These authors contributed equally to this work.

Abstract: During the production process of commercial carbon fiber reinforced polymers (CFRPs), a silane coupling agent is added to the carbon fiber at the sizing step as a binder to enhance the product's physical properties. While improving strength, the silane coupling agent results in a silane residue on recovered carbon fibers (rCF) after recycling, which is a disadvantage when using recovered carbon fibers in the manufacture of new materials. In this study, the rCF is recovered from waste carbon fiber reinforced polymers (CFRPs) from the bicycle industry by a microwave pyrolysis method, applying a short reaction time and in an air atmosphere. Moreover, the rCF are investigated for their surface morphologies and the elements present on the surface. The silicon element content changes with pyrolysis temperature were 0.4, 0.9, and 0.2%, respectively, at 450, 550, and 650 °C. Additionally, at 950 °C, silicon content can be reduced to 0.1 ± 0.05%. The uniformity of microwave pyrolysis recycle treatment was compared with traditional furnace techniques used for bulk waste treatment by applying the same temperature regime. This work provides evidence that microwave pyrolysis can be used as an alternative method for the production of rCFs for reuse applications.

Keywords: carbon fiber-reinforced polymer; recycled carbon fiber; microwave-assisted pyrolysis; silane coupling agents

Citation: Chin, K.-Y.; Shiue, A.; Wu, Y.-J.; Chang, S.-M.; Li, Y.-F.; Shen, M.-Y.; Leggett, G. Studies on Recycling Silane Controllable Recovered Carbon Fiber from Waste CFRP. *Sustainability* **2022**, *14*, 700. https://doi.org/10.3390/su14020700

Academic Editor: Antonio Caggiano

Received: 28 October 2021
Accepted: 6 January 2022
Published: 9 January 2022

1. Introduction

In recent decades, carbon fiber-reinforced polymers (CFRPs) have been extensively utilized in various applications of the aerospace industry, transportation, infrastructure [1–3], energy, sport industries, defense, medical sector, and electronics [4,5]. The physical properties of CFRPs have led to increased use of these materials as replacements for more conventional options, including steel, aluminum, alloys, etc. These physical properties include low density and lightweight, superior strength to weight ratios and elastic modulus [6–8], stiffness, low expansion or shrinkage, thermal stability [9–12], electrical conductivity, superior resistance to corrosion and chemical attack, and exceptional endurance of physico–chemical properties [13–16]. Given these properties, they have been heavily utilized in the manufacture of automotive parts to reduce weight and improve fuel efficiency [17]. CFRPs' composites are engineered materials that compose of carbon fibers (CFs) acting as reinforcing matrix materials within versed thermosetting polymer to form structures. [18]. The fibers act as load-carrying elements for their orientation and to resist environmental damage [11,19].

CFRPs' composite materials are differentiated by silane with different properties, suitable for a variety of strength requirements. In response to increased demand to reduce CO_2 emissions through weight saving measures and production related energy costs, the use of CFRP products and applications has a high potential to grow significantly over the coming years. Carbon fibers were first commercially produced in the late 1960s. They are now manufactured around the world. It is predicted that by 2025, revenues associated with use of CFRPs will exceed 25 billion dollars per year [20,21]. Recent annual growth rates of more than 10% underline these forecasts [22]. Given that the service life of CFRPs is approximately 50 years, as end of life is reached, this will bring about a considerable increase in CFRPs waste generation and present a significant challenge for the disposal and recycling of CFRPs based components [6,23].

With increasing global focus on a circular economy model at domestic and foreign scale, the problems of recycling and waste reduction are becoming more and more prominent, with industry increasingly considering the impact of growth on finite resources. The presence of carbon fibers in composite materials is a challenge when it comes to recycling and separating the fibrous material from epoxy resins, plastics, and other materials. In recent years, industries have speedily adopted these materials. However, it has displayed no proper awareness and consideration of their disposal and possible recycling. The main treatment methods for waste CFRPs have been landfill disposal and incineration [9,24]. Economic viability, new legislations, limited landfill capacity, and fuel incineration, and ecological aspects are driving towards recycling and processing for CFRPs waste [25,26]. Thus, the treatment methodology of CFRP waste is becoming a more critical issue, with industry currently lacking in the skills and methodologies to respond effectively. Support is required to provide industry with tools that can be implemented effectively and economically.

Currently, several technologies have been developed for recovering CFs from CFRPs with excellent recycling yield and maintenance of critical physical properties. There are three main approaches:

(1) Mechanical processes: shredding, crushing and other physical methods effect separation of CFs and raw matrix materials [2,27,28], or high voltage pressure (electrodynamic fragmentation) [29,30]. The recovery cost is low and the method is simple. Production of pollutants is minimal. However, the damage to CFs in the recovery process is high, making the recovered material unsuitable for the manufacture of typical CF-based products. As such, material recovered in this way is usually made into powder for filling or as a reinforcing material.

(2) Chemical recovery: separate CFs from the matrix resin [2,9,31–38]. The damage to recovered CFs is minimal, but the cost of chemical agents is high and this technology is still very much in its infancy. Waste solutions from the process require recovery and subsequent treatment before reuse. In the absence of this step, the result is the production of secondary pollution of large volumes of liquid waste and hazardous gases, the disposal of which presents a significant hazard to the environment [6]. In addition, the method does not allow simple recovery of CFs in large quantities. At present, this method is rarely used.

(3) Thermal technology: The matrix material is decomposed by heat (combustion/incineration, fluidized bed process, or pyrolysis) [2,7,39–42] or microwave radiation [43,44]. Heat is utilized to decompose the scrap composite in the thermal recycling process. Due to a higher operating temperature at ~450–700 °C [45–48], combustible materials are burnt, leaving the CF material for recovery. Generally, the process temperature is dependent on the type of resin used in the scrap composite. Irregular temperature can leave char on the undercooked fiber surface or result in a decrease in the diameter of the overheated recovered fibers [49,50].

Microwave pyrolysis is a relatively new approach, designed to replace conventional pyrolysis. Heating by microwave pyrolysis has the advantage that the rate of thermal transfer is increased, with reduced energy consumption [26,51] and heat loss reduction [52,53]. This process can recycle CFRPs, with resulting fibers maintaining desirable mechanical properties [49,54–58] These studies are focused on removing resin and maintaining mechanical strength to get high quality rCF, without considering the issue of residual silane on the recovered carbon fibers. However, the manufacturing process of commercial CFRPs involves adding an appropriate amount of silane [59,60]. Silanes are often used as a coupling agent, possessing differing functional groups at each molecular end. For example, trialkoxysilane can bond with inorganic substances. Alternatively, there are organofunctional groups (methacrylate, epoxy, etc.). These silanes can enhance the compatibility between carbon fibers and the resin matrix, improving the interfacial bonding strength of CFRPs. Therefore, if silane could be saved during the fiber recovery process it would reduce the need for the addition of further silane coupling agent when using rCF to manufacture new materials, reducing overall costs and waste [61]. Furthermore, Li et al. [62] compare the dispersion of sizing of carbon fiber after removed silane treatment and nontreatment, with the product intended to be added to portland cement. The results show the carbon fiber of removed silane can be efficiently dispersed to form carbon-fiber-reinforced cement, with this composite cement exhibiting higher compressive strength. The residue of silane could be a disadvantage when considering recovered carbon fibers for reinforced inorganic applications.

In this study, the residual silane content of rCF surface can be controlled by the temperature of microwave pyrolysis. The influence of the reaction temperature on the degradation of silicon on the carbon fiber surface was studied. The recovered carbon fibers were characterized by scanning electron microscopy (SEM), thermogravimetric analysis (TGA), and Energy-dispersive X-ray spectroscopy (EDS).

2. Experimental Methods

2.1. Materials

The carbon fiber reinforced composite material (CFRPs) waste from recycled bicycle frame and the original Carbon fiber (Mitsubishi TR 50) was supplied by Giant Manufacturing Co., Ltd (Taichung City, Taiwan). The waste was cut and crushed by plate-cutting and hydraulic press machine to the sample size of about 90 mm × 90 mm × 4.5 mm.

2.2. Pyrolysis

The traditional pyrolysis process is performed using a typically muffle furnace (JH-5, Taiwan, 1200 W). The CFRP waste (10 g) was loaded into a ceramic crucible and pyrolyzed at 650 °C in an air atmosphere. Figure 1 shows the micro-wave pyrolysis system (MILESTONE PYRO 260). Pyrolysis of the chips of waste (10 g) was carried out at 350–950 °C in an air atmosphere. The special microwave–transparent ceramic muffle furnace allows microwave energy to transmit through and rapidly raise the temperature of silicon carbide plates situated inside the muffle. Microwaves were generated by a single magnetron system with rotating diffuser for homogeneous microwave distribution, the power was set up to 1200 W and the frequency of the microwave was set to 2.45 GHz. The detail condition shows at Table 1.

2.3. Characterization

The CFRPs wastes are characterized using Thermogravimetric Analysis (TGA), Scanning Electron Microscopy images and Energy-dispersive X-ray spectroscopy (SEM-EDS). Netzsch instrument TG 209 F3 model is used to make TGA measurements. The measurements were characterized for the thermal degradation curve of the CFRP characterized under an air atmosphere. The weight of the sample is 10 mg. Each sample was heated from room temperature to 650 °C at a heating rate of 10 °C/min. Three target areas were identified for each of the rCF samples, and taking the core layer of each sample, the surface element (C, O, and Si) of the core layer is measured using a Hitachi TM 4000 Plus instrument with Oxford AZtecOne (EDS) to assess removal capability and uniformity for silane and resin. The high magnification physical morphology of carbon fiber was detected by Hitachi Regulus 8100 field-emission scanning electron microscopes (FE-SEM).

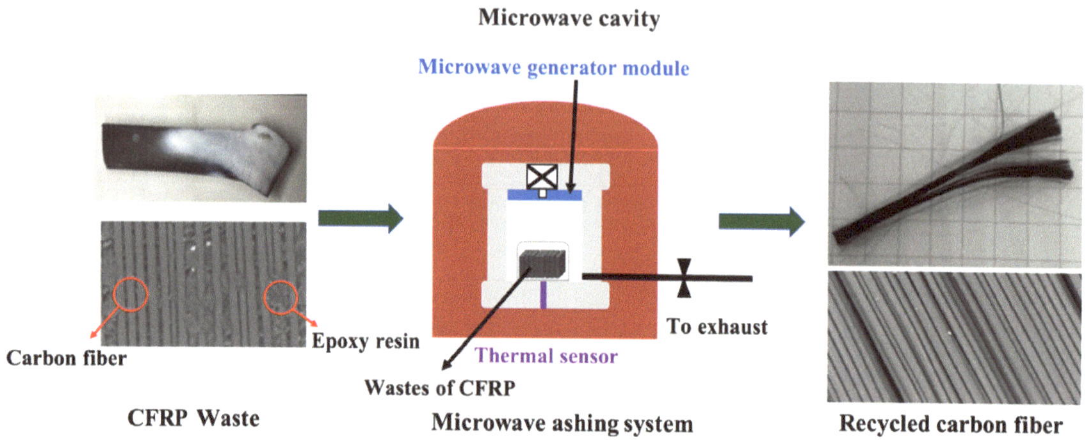

Figure 1. CFRPs waste recycling treatment by micro-wave pyrolysis system.

Table 1. Details of samples at different reaction conditions.

Samples	Temperature/(°C) *	Average Power/W	Heating Method
TF 650 °C	650	1200	Traditional
MW 350 °C	350	1200	Microwave
MW 450 °C	450	1200	Microwave
MW 550 °C	550	1200	Microwave
MW 650 °C	650	1200	Microwave
MW 950 °C	950	1200	Microwave

* There was no retention time at the desired temperature.

3. Experimental Results and Discussions

3.1. TGA Analysis of CFRPs Waste

It is important to understand the thermal decomposition curve of the CFRPs' waste from bicycle frame waste, since cause each batch of CFRP could be from different source. As shown in Figure 2, the DTG curve has a significant mass loss in the range of 250 to 450 °C and 490 to 560 °C, respectively. Therefore, it can be defined that the decomposition temperature of the first TGA stage is 220 to 450 °C and the second stage is 475 to 550 °C. The third stage of the decomposition range is between 600 and 650 °C. The thermal decomposition curve is consistent with results presented by Yatim et al. [57] and Deng et al. [55].

In Figure 2B–D, the SEM image also shows the extent to which CFRP waste is removed from the resin between the fibers of the CFRPs' waste at various stages of TGA. In stage 1, at 330 °C (Figure 2B), a weight loss of approximately 5% occurred. It can be observed that there is still a large amount of epoxy residue between the fibers. In stage 2 (Figure 2C), after heating beyond the first interval, the weight loss is 30%. The gaps between the fibers can be clearly seen and there is still un-removeable resin around the carbon fiber. After 500 to 600 °C, second interval, an increasing amount of resin is removed; approximately 20%. No residual resin was observed on the surface of carbon fiber. Based on TGA thermal weight loss results, it can be speculated that the carbon fiber-to-resin weight ratio of the CFRPs waste is approximately 40:60.

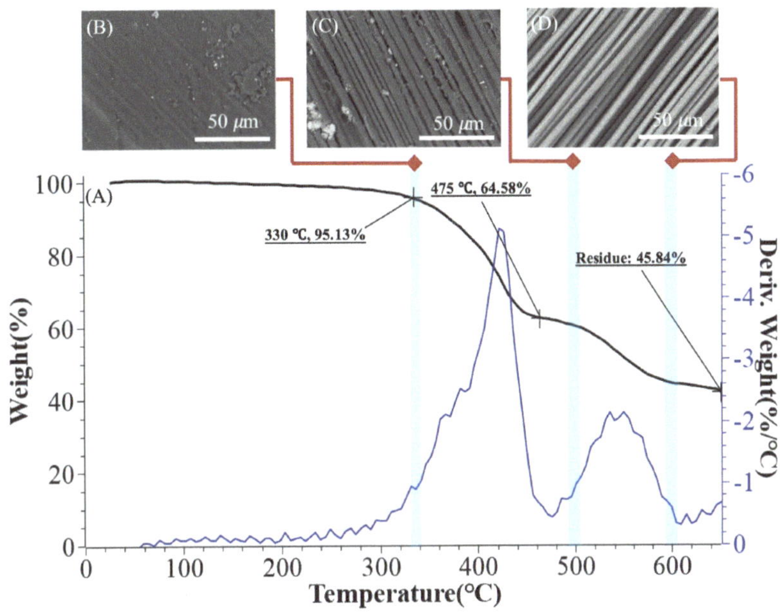

Figure 2. TGA analysis of CFRP waste: (**A**) CFRP of thermal decomposition curve; and (**B–D**) SEM micrograph of CRFP thermal decomposition stages (**B**) stage 1, (**C**) stage 2, and (**D**) stage 3.

3.2. Compare the Pyrolysis Result of Microwave with Traditional Treatment for CFRPs Waste

The parameters obtained from TGA thermal decomposition were used as the reference condition for the removal of the resin by high temperature furnace and microwave pyrolysis. Figure 3 presents CFRP waste samples recovered using two different heat treatment methods. Figure 3A shows SEM image and EDS analysis results of virgin waste carbon fiber. The SEM image clearly shows a significant quantity of resin wrapped around the carbon fiber. EDS elemental analysis shows the presence of carbon, oxygen, and silicon content at 85.8 ± 5.3%, 13.6 ± 4.7%, and 0.3 ± 0.6%, respectively. Figure 3B shows no significant difference between CFRP waste in high-temperature furnaces after heating up to 650 °C, except for a color change resulting in a darker appearance. And in Figure 3C, CFRP waste decomposed by microwave pyrolysis results in carbon fibers that can be separated easily into individual fibers.

3.3. Influence of Different Temperature on Removed Resin in Large Scale Waste CFRP for Microwave Pyrolysis

In this study, the samples of waste CFRPs were obtained from recycled bicycle frames. Samples were prepared to the required sizes through smash and cut methods; the thickness being approximately 4.5 mm. The CFRP was closely composed of multilayer CF and resin. The pyrolysis of the resin requires enough oxygen and heat for carbon oxidation to occur and for effective removal of resin from CFRP wastes. The irregular and curled shape of CFRP wastes and the compact multilayer CF structure could easily result in non-uniform hot spots in the inside of each waste element and this structure also makes it difficult for the resin to make good contact with surrounding air. Therefore, the resin removal methodology requires further research [56].

To understand the uniformity of waste CFRP at large scales under microwave assisted pyrolysis, three parts of the samples in the crucible were studied via SEM and EDS as shown in Figure 4 (marked 1, 2, and 3). The effectiveness of resin removal was assessed through SEM images and with carbon and oxygen content determined using EDS element analysis. Each RCF sample was peeled off from the outside, and its core used for SEM/EDS measurement. EDS analysis results of the samples are presented in Table 2.

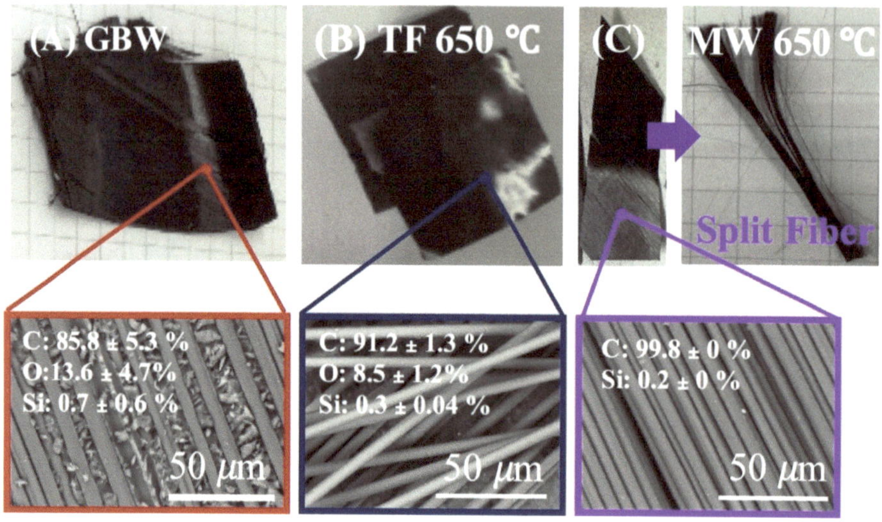

Figure 3. Optical photo, SEM imagine, and EDS analysis: (**A**) CFRP waste of Giant Bicycles (GBW), (**B**) CFRPs waste sample (TF 650 °C) after treatment with traditional furnace pyrolysis at 650 °C, and (**C**) CFRPs waste (MW 650 °C) after treatment with micro-wave pyrolysis at 650 °C.

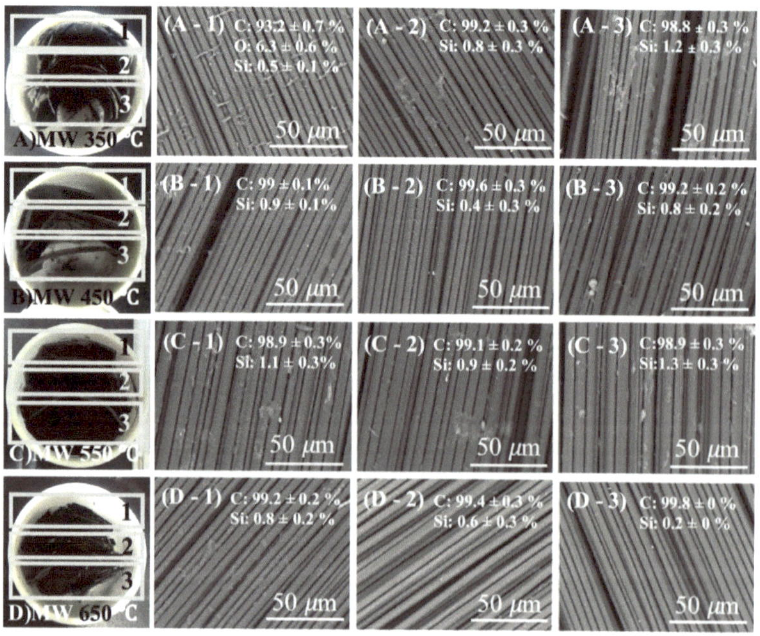

Figure 4. Photograph and SEM micrographs of the recycled carbon fibers in different part of ceramic crucible after microwave pyrolysis with varying condition; (**A**) microwave pyrolysis at 350 °C, (**B**) at 450 °C, (**C**) at 550 °C and (**D**) at 650 °C.

Table 2. EDS analysis of the samples.

		Element% by Mass [a]			Average of Area 1,2,3
Samples	**Area**	**C**	**O**	**Si**	**Si% by Mass**
GBW	-	85.8 ± 5.3	13.6 ± 4.7	0.7 ± 0.6	-
TF 650 °C	-	91.2 ± 1.3	8.5 ± 1.2	0.3 ± 0.4	-
MW 350 °C	1	93.2 ± 0.7	6.3 ± 0.6	0.5 ± 0.1	0.8 ± 0.3
	2	99.2 ± 0.3	-	0.8 ± 0.3	
	3	98.8 ± 0.3	-	1.2 ± 0.3	
MW 450 °C	1	99 ± 0.1	-	0.9 ± 0.1	0.7 ± 0.3
	2	99.6 ± 0.3	-	0.4 ± 0.3	
	3	99.2 ± 0.2	-	0.8 ± 0.2	
MW 550 °C	1	98.9 ± 0.3	-	1.1 ± 0.3	1.1 ± 0.3
	2	99.1 ± 0.2	-	0.9 ± 0.2	
	3	98.9 ± 0.3	-	1.3 ± 0.3	
MW 650 °C	1	99.2 ± 0.2	-	0.8 ± 0.2	0.5 ± 0.3
	2	99.4 ± 0.3	-	0.6 ± 0.3	
	3	99.8 ±0.1	-	0.2 ± 0.1	
MW 950 °C	-	99.9 ±0.1	-	0.1 ± 0.1	-

[a] Each value is an average taken over 6 different spots on the samples.

At 350 °C pyrolysis temperature (Figure 4(A-1–A-3)), the A-1 area of SEM image shows the resin residue between the carbon fibers. EDS analysis determined the carbon, oxygen, and silicon content as 93.2 ± 0.7, 6.3 ± 0.6, and 0.5 ± 0.1%, respectively. Compared with samples GBW of Figure 3A, oxygen content decreased from 13.6 ± 4.7 to 6.3 ± 0.6% and carbon content increased from 85.8 ± 5.3 to 93.2 ± 0.7%. These results revealed that the resin is not fully pyrolyzed. In addition, the EDS and SEM results of A-2 and A-3 areas are similar. These SEM images show a significant quantity of resin adhered to the carbon fiber. Comparing the EDS results of three areas (Figure 4(A-1–A-3), Table 1), areas 2 and 3 show better resin removal. A similar situation also occurred under the different sample pyrolysis conditions applied.

At 450 °C (Figure 4(B-1–B-3), Table 1), the EDS results showed the carbon content is 99 ± 0.1, 99.6 ± 0.3, and 99.2 ± 0.2% and silicon content is 0.9 ± 0.1, 0.4 ± 0.3, and 0.8 ± 0.2%, respectively. Figure 4C,(D-1–D-3), shows the EDS and SEM results after microwave pyrolysis at 550 °C (B) and 650 °C (D). The carbon content of MW 550 °C from areas 1 to 3 is 98.9 ± 0.3, 99.1 ± 0.2, and 98.9 ± 0.3% and silicon content is 1.1 ± 0.3, 0.9 ± 0.2, and 1.3 0.3%, respectively. For MW650 °C, the carbon content is 99.2 ± 0.2, 99.4 ± 0.3, and 99.8 ± 0%, and the silicon content is 0.80.2, 0.6 ± 0.3 and 0.2 ± 0%, respectively. All of the above EDS results show that pyrolysis of waste CFRP is uniform at different crucible positions. While the pyrolysis condition is above 450 °C, the oxygen content reduced to 0%, indicating that the resin was fully carbonized. The difference between carbon content for the three areas is approximately 0.47 ± 0.01% at 550 and 650 °C. This was, perhaps, caused by the waste having a curved and irregular shape or non-uniform density. Verification of these statements requires further research. However, this pyrolysis condition had enough process time and temperature to reduce the influence caused by different crucible position.

The SEM images show filamentous matter on the fibers. The shape changed from the filament to a dot along with decreased silicon content. Section 3.4 will discuss silicon removal from CFRP waste through microwave pyrolysis.

3.4. Silane Removal from CFRPs Waste at Different Microwave Pyrolysis Temperature

Figure 5 shows a SEM diagram of CFRP materials that are thermally treated at different temperatures. A carbon fiber sample (pyrolysis 650) treated at 650 °C in a high-temperature oven shows significant resin filling, as shown in Figure 5A. The results of the EDS element analysis of the sample showed that the fiber surface of TF at 650 °C had an oxygen content of approximately 8.5 ± 1.2%. Based on results of the TGA analysis, the resin had been carbonized to form carbide; this means, for a traditional furnace, resin pyrolysis is not complete. For the residual resin layer wrapped on the surface of waste CFRPs, the Si element content is 0.3 ± 0.04%, while the CFRPs samples treated with microwave pyrolysis at 350 °C (Figure 5B), 450 °C (Figure 5C), and 550 °C (Figure 5D), show a cotton-like or filamentous substance on the surface of the carbon fibers. The silicon content is approximately 0.9 to 0.4%. At a treatment temperature of 650 °C (Figure 5E), the shape of filamentous substances changed to dots (silicon content: 0.2 ± 0%).

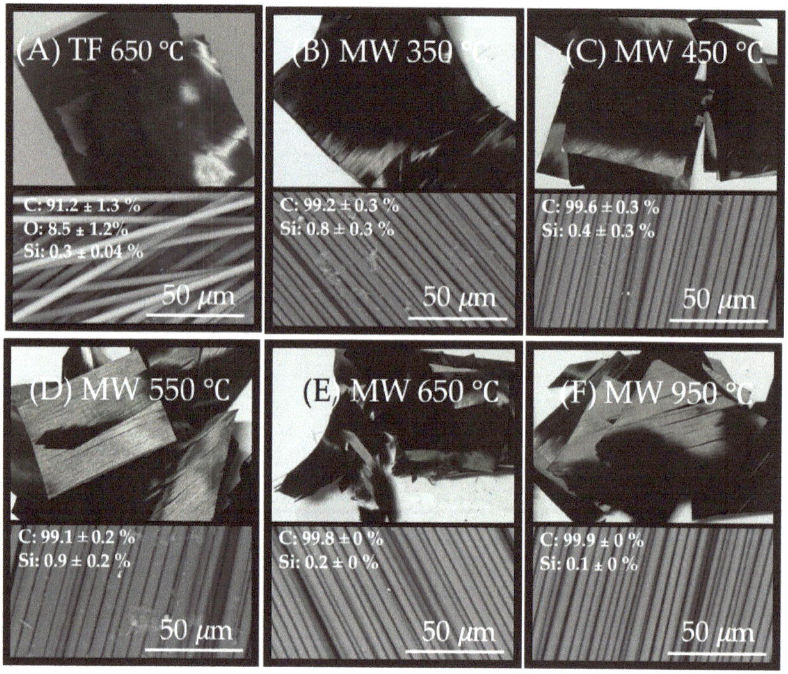

Figure 5. SEM micrographs of the recycled carbon fibers by thermally treated at different temperatures; (**A**) traditional furnace pyrolysis treated at 650 °C, (**B**) microwave pyrolysis treated at 350 °C, (**C**) at 450 °C, (**D**) at 550 °C, (**E**) at 650 °C; and (**F**) 950 °C.

By way of understanding the effect of microwave thermal pyrolysis temperature on residual silane content on waste CFRPs surface, we also set the temperature of microwave pyrolysis at 950 °C. Following this treatment, the silicon content is 0.1%. The SEM image shows that the recovered carbon fiber surface is completely free of residues or obvious defects, as shown in Figure 5F. In Figure 6, the MW 950 °C Figure 6A,B compared with the original carbon fiber (Mitsubishi TR 50, Figure 6C,D), demonstrates they are without defects on the fiber surface, which can be further proved by high magnification FE-SEM.

However, even if the microwave pyrolysis temperature increases by 300 °C, the silicon content is only slightly reduced. Therefore, the results show the silicon content is gradually reduced from the recovered carbon fiber along with breakdown of the resins, but it cannot be completely removed by microwave pyrolysis at 350 to 950 °C.

Since the organic components in the silane compounds typically degrade when heated to approximately 400 °C, only elemental silicon remains. Silicon gasification temperature is 3265 °C, which is difficult to achieve via thermal decomposition methods. Other researchers also presented SEM/EDS results of pyrolyzed carbon fiber pretreated using silane coupling agent that exhibit the same dots on the carbon fiber surface and were determined as containing elemental silicon [63,64].

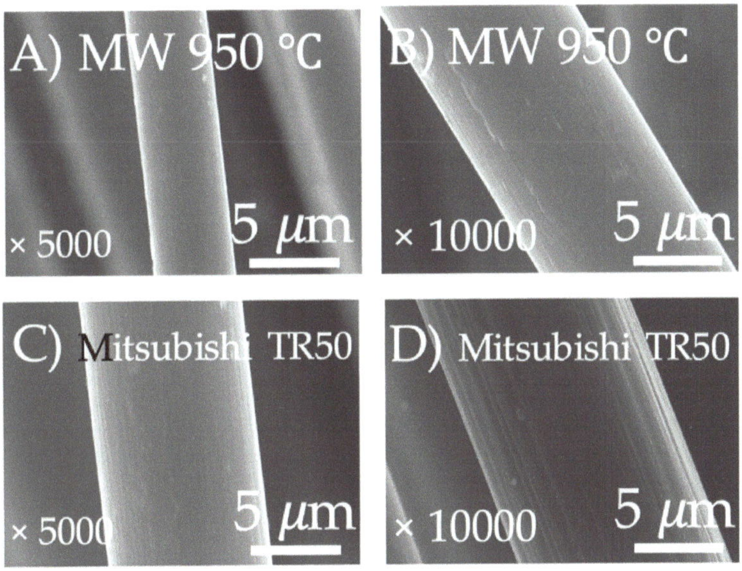

Figure 6. High magnification FE-SEM micrographs of the recycled carbon fibers (MW 950 °C, (**A**); at 5000× magnification, (**B**); 10,000×) and the original carbon fiber (Mitsubishi TR 50, (**C**,**D**) they magnification ware respectively 5000× and 10,000×).

For these reasons, these filamentous substances are inferred to be residual silicon and carbonized resin produced during the CFRPs pyrolysis process. The residual silicon is removed as the carbonized resin pyrolyzed to form carbon dioxide or carbon monoxide by air atmosphere. Despite the evidence for the above conclusions, further research is required to affirm this interpretation.

4. Conclusions

The silane controllable recycled carbon fiber can effectively simplify the process of rCF reuse in suitable applications. Through adjusted silane content, the dispersion of rCF in organic or inorganic materials was improved for use in the manufacture of high-performance composite materials.

This study successfully demonstrates control of the silane content on the rCF surface. Different microwave pyrolysis methods and temperature conditions were compared. The results show that the microwave pyrolysis method achieves better resin removal when compared with traditional thermal treatment methods at the same temperature. Over 450 °C, the SEM/EDS results show the recovered surface of the carbon fibers are absent of oxygen. Compared with the GBW sample, no additional defects were observed. Moreover, the resin and silane removal uniformities were examined from recovered carbon fiber in large scale wastes. At different crucible position, the difference of maximum silane content is about ±0.33%. Moreover, a higher treatment temperature can reduce this difference.

The results during microwave pyrolysis could be reproduced for this CFRP waste of Giant Bicycles discussed in this article. However, a multitude of influencing parameters affect the reclamation of carbon fibers, such as matrix material, temperatures, atmosphere composition, etc. This complicates the recycling process control to reclaim high quality fibers. Further research into the decomposition behavior of CFRPs recovered from other, commonly recycled industrial composite materials is required.

Author Contributions: ConceptualizationConceptualization, K.-Y.C. and S.-M.C.; data curation, K.-Y.C. and Y.-J.W.; formal analysis, K.-Y.C. and Y.-J.W.; investigation, K.-Y.C. and A.S.; methodology, K.-Y.C., Y.-J.W. and S.-M.C.; project administration, K.-Y.C. and Y.-J.W.; supervision, A.S., S.-M.C., Y.-F.L. and M.-Y.S.; writing—original draft, K.-Y.C. and A.S.; writing—review and editing, A.S., G.L., Y.-F.L., S.-M.C. and M.-Y.S. All authors have read and agreed to the published version of the manuscript.

Funding: This research was funded by the Ministry of Science and Technology of Taiwan (108-2218-E-027-005, 109-2221-E-027-113), the Ministry of Education in Taiwan, and the Topkey Corp. Taiwan.

Institutional Review Board Statement: Not applicable.

Informed Consent Statement: Not applicable.

Data Availability Statement: Not applicable.

Conflicts of Interest: The authors declare no conflict of interest.

References

1. Asmatulu, E.; Twomey, J.M.; Overcash, M. Recycling of fiber-reinforced composites and direct structural composite recycling concept. *J. Compos. Mater.* **2013**, *48*, 593–608. [CrossRef]
2. Oliveux, G.; Dandy, L.O.; Leeke, G. Current status of recycling of fibre reinforced polymers: Review of technologies, reuse and resulting properties. *Prog. Mater. Sci.* **2015**, *72*, 61–99. [CrossRef]
3. Jensen, J.; Skelton, K. Wind turbine blade recycling: Experiences, challenges and possibilities in a circular economy. *Renew. Sustain. Energy Rev.* **2018**, *97*, 165–176. [CrossRef]
4. Pimenta, S.; Pinho, S. Recycling carbon fibre reinforced polymers for structural applications: Technology review and market outlook. *Waste Manag.* **2011**, *31*, 378–392. [CrossRef]
5. Cousins, D.S.; Suzuki, Y.; Murray, R.; Samaniuk, J.R.; Stebner, A.P. Recycling glass fiber thermoplastic composites from wind turbine blades. *J. Clean. Prod.* **2018**, *209*, 1252–1263. [CrossRef]
6. Sun, H.; Guo, G.; Memon, S.A.; Xu, W.; Zhang, Q.; Zhu, J.-H.; Xing, F. Recycling of carbon fibers from carbon fiber reinforced polymer using electrochemical method. *Compos. Part A Appl. Sci. Manuf.* **2015**, *78*, 10–17. [CrossRef]
7. Oliveux, G.; Bailleul, J.-L.; Gillet, A.; Mantaux, O.; Leeke, G. Recovery and reuse of discontinuous carbon fibres by solvolysis: Realignment and properties of remanufactured materials. *Compos. Sci. Technol.* **2017**, *139*, 99–108. [CrossRef]
8. Yang, J.; Liu, J.; Liu, W.; Wang, J.; Tang, T. Recycling of carbon fibre reinforced epoxy resin composites under various oxygen concentrations in nitrogen–oxygen atmosphere. *J. Anal. Appl. Pyrolysis* **2015**, *112*, 253–261. [CrossRef]
9. Xu, P.; Li, J.; Ding, J. Chemical recycling of carbon fibre/epoxy composites in a mixed solution of peroxide hydrogen and N,N-dimethylformamide. *Compos. Sci. Technol.* **2013**, *82*, 54–59. [CrossRef]
10. Frohs, W.; Jaeger, H. Carbon fiber & composite material—Landscape Germany. *Carbon* **2012**, *50*, 737.
11. Yao, S.S.; Jin, F.L.; Rhee, K.Y.; Hui, D.; Park, S.J. Recent advances in carbonfiber-reinforced thermoplastic composites: A review. *Compos. Part B Eng.* **2018**, *142*, 241–250. [CrossRef]
12. Bachmann, J.; Hidalgo, C.; Bricout, S. Environmental analysis of innovative sustainable composites with potential use in avia-tion sector—A life cycle assessment review. *Sci. China Technol. Sci.* **2017**, *60*, 1301–1317. [CrossRef]
13. Zhu, J.-H.; Wei, L.; Moahmoud, H.; Redaelli, E.; Xing, F.; Bertolini, L. Investigation on CFRP as dual-functional material in chloride-contaminated solutions. *Constr. Build. Mater.* **2017**, *151*, 127–137. [CrossRef]
14. Zhu, J.-H.; Su, M.-N.; Huang, J.-Y.; Ueda, T.; Xing, F. The ICCP-SS technique for retrofitting reinforced concrete compressive members subjected to corrosion. *Constr. Build. Mater.* **2018**, *167*, 669–679. [CrossRef]
15. Yu, K.; Shi, Q.; Dunn, M.L.; Wang, T.T.; Qi, H.J. Carbon Fiber Reinforced Thermoset Composite with Near 100% Recyclability. *Adv. Funct. Mater.* **2016**, *26*, 6098. [CrossRef]
16. Hadigheh, S.; Kashi, S. Effectiveness of vacuum consolidation in bonding fibre reinforced polymer (FRP) composites onto concrete surfaces. *Constr. Build. Mater.* **2018**, *187*, 854–864. [CrossRef]
17. Yoshinori Kakichi, Y.; Yamaguchi, A.; Hashimoto, T.; Urushisaki, M.; Sakaguchi, T.; Kawabe, K.; Kondo, K.; Iyo, H. Development of recyclable carbon fiber-reinforced plastics (CFRPs) with controlled degradability and stability using acetal link-age-containing epoxy resins. *Polym. J.* **2017**, *49*, 851–859. [CrossRef]
18. Kalpakjian, S.; Schmid, S.R. *Manufacturing Engineering and Technology*, 4th ed.; Prentice Hall: Hoboken, NJ, USA, 2000.
19. Lau, K.-T.; Hung, P.-Y.; Zhu, M.-H.; Hui, D. Properties of natural fibre composites for structural engineering applications. *Compos. Part B Eng.* **2018**, *136*, 222–233. [CrossRef]
20. Kraus, T.; Kühnel, M.; Witten, E. *Composites Market Report 2014Market Developments, Trends, Challenges and Opportunities*; Federation of Reinforced Plastics: Frankfurt, Germany, 2014.
21. Roberts, T. *The Carbon Fibre Industry: Global Strategic Market Evaluation 2006–2010*; Materials Technology Publications: Watford, UK, 2006; Volume 10.

22. Krauklis, A.E.; Karl, C.W.; Gagani, A.I.; Jørgensen, J.K. Composite material recycling technology—State-of-the-art and sustainable development for the 2020s. *J. Compos. Sci.* **2021**, *5*, 28. [CrossRef]
23. Jiang, G.; Pickering, S.; Walker, G.; Wong, K.; Rudd, C. Surface characterisation of carbon fibre recycled using fluidised bed. *Appl. Surf. Sci.* **2007**, *254*, 2588–2593. [CrossRef]
24. Witik, R.A.; Teuscher, R.; Michaud, V.; Ludwig, C.; Månson, J.A.E. Carbon fibre reinforced composite waste: An environmental assessment of recycling, energy recovery and landfilling. *Compos. Part A Appl. Sci. Manuf.* **2013**, *49*, 89–99. [CrossRef]
25. Meng, F. Environmental and Cost Analysis of Carbon Fibre Composites Recycling. Ph.D. Thesis, University of Nottingham, Nottingham, UK, 2017.
26. Gopalraj, S.K.; Kärki, T. A review on the recycling of waste carbon fibre/glass fibre-reinforced composites: Fibre recovery, properties and life-cycle analysis. *SN Appl. Sci.* **2020**, *2*, 1–21. [CrossRef]
27. Kouparitsas, C.E.; Kartalis, C.N.; Varelidis, P.C.; Tsenoglou, C.J.; Papaspyrides, C.D. Recycling of the fibrous fraction of rein-forced thermoset composites. *Polym. Compos.* **2010**, *23*, 682–689. [CrossRef]
28. Howarth, J.; Mareddy, S.S.; Mativenga, P. Energy intensity and environmental analysis of mechanical recycling of carbon fibre composite. *J. Clean. Prod.* **2014**, *81*, 46–50. [CrossRef]
29. Vo Dong, P.A.; Azzaro-Pantel, C.; Cadene, A.L. Economic and environmental assessment of recovery and disposal pathways for CFRP waste management. *Resour. Conserv. Recycl.* **2018**, *133*, 63–75. [CrossRef]
30. Mativenga, P.; Shuaib, N.A.; Howarth, J.; Pestalozzi, F.; Woidasky, J. High voltage fragmentation and mechanical recycling of glass fibre thermoset composite. *CIRP Ann.* **2016**, *65*, 45–48. [CrossRef]
31. Bai, Y.; Wang, Z.; Feng, L. Chemical recycling of carbon fibers reinforced epoxy resin composites in oxygen in supercritical water. *Mater. Des.* **2010**, *31*, 999–1002. [CrossRef]
32. Liu, Y.; Liu, J.; Jiang, Z.; Tang, T. Chemical recycling of carbon fibre reinforced epoxy resin composites in subcritical water: Synergistic e_ect of phenol and koh on the decomposition efficiency. *Polym. Degrad. Stab.* **2012**, *97*, 214–220. [CrossRef]
33. Yan, H.; Lu, C.-X.; Jing, D.-Q.; Chang, C.-B.; Liu, N.-X.; Hou, X.-L. Recycling of carbon fibers in epoxy resin composites using supercritical 1-propanol. *Carbon* **2016**, *100*, 710–711. [CrossRef]
34. Yildirir, E.; Onwudili, J.A.; Williams, P.T. Recovery of carbon fibres and production of high quality fuel gas from the chemical recycling of carbon fibre reinforced plastic wastes. *J. Supercrit. Fluids* **2014**, *92*, 107–114. [CrossRef]
35. Oshima, K.; Fujii, H.; Morita, K.; Hosaka, M.; Muroi, T.; Satokawa, S. Selective Phenol Recovery by Catalytic Cracking of Thermal Decomposition Gas from Epoxy-Based Carbon-Fiber-Reinforced Plastic. *Ind. Eng. Chem. Res.* **2020**, *59*, 13460–13466. [CrossRef]
36. Lo, J.N.; Nutt, S.R.; Williams, T.J. Recycling Benzoxazine–Epoxy Composites via Catalytic Oxidation. *ACS Sustain. Chem. Eng.* **2018**, *6*, 7227–7231. [CrossRef]
37. Das, M.; Chacko, R.; Varughese, S. An Efficient Method of Recycling of CFRP Waste Using Peracetic Acid. *ACS Sustain. Chem. Eng.* **2018**, *6*, 1564–1571. [CrossRef]
38. Borjan, D.; Knez, Ž.; Knez, M. Recycling of carbon fiber-reinforced composites—Difficulties and future perspectives. *Materials* **2021**, *14*, 4191. [CrossRef] [PubMed]
39. López, F.; Martín, M.; Alguacil, F.; Rincón, J.M.; Centeno, T.; Romero, M. Thermolysis of fibreglass polyester composite and reutilisation of the glass fibre residue to obtain a glass–ceramic material. *J. Anal. Appl. Pyrolysis* **2012**, *93*, 104–112. [CrossRef]
40. López, F.A.; Rodríguez, O.; Alguacil, F.J.; García-Díaz, I.; Centeno, T.A.; García-Fierro, J.L.; González, C. Recovery of carbon fibres by the thermolysis and gasification of waste prepreg. *J. Anal. Appl. Pyrolysis* **2013**, *104*, 675–683. [CrossRef]
41. Yip, H.L.H.; Pickering, S.; Rudd, C. Characterisation of carbon fibres recycled from scrap composites using fluidised bed process. *Plast. Rubber Compos.* **2002**, *31*, 278–282. [CrossRef]
42. Ye, S.Y.; Bounaceur, A.; Soudais, Y.; Barna, R. Parameter Optimization of the Steam Thermolysis: A Process to Recover Carbon Fibers from Polymer-Matrix Composites. *Waste Biomass Valorization* **2013**, *4*, 73–86. [CrossRef]
43. Akesson, D.; Foltynowicz, Z.; Christeen, J.; Skrifvars, M. Products obtained from decomposition of glass fibre-reinforced composites using microwave pyrolysis. *Polimery* **2013**, *58*, 582–586. [CrossRef]
44. Obunai, K.; Fukuta, T.; Ozaki, K. Carbon fiber extraction from waste CFRP by microwave irradiation. *Compos. Part A Appl. Sci. Manuf.* **2015**, *78*, 160–165. [CrossRef]
45. De Moraes, V.T.; Jermolovicius, L.A.; Tenório, J.A.S.; Lebrão, S.M.G.; Lebrãoa, G.W. Microwave-assisted recycling process to recover fiber from fiberglass polyester composites. *Mater. Res.* **2019**, *22*, 1–8. [CrossRef]
46. Limburg, M.; Stockschläder, J.; Quicker, P. Thermal treatment of carbon fibre reinforced polymers (Part 1: Recycling). *Waste Manag. Res.* **2019**, *37*, 73–82. [CrossRef]
47. Yatim, N.M.; Shamsudin, Z.; Shaaban, A.; Sani, N.A.; Jumaidin, R.; Shariff, E.A. Thermal analysis of carbon fibre reinforced polymer decomposition. *Mater. Res. Express* **2020**, *7*, 015615. [CrossRef]
48. Naqvi, S.R.; Prabhakara, H.M.; Bramer, E.; Dierkes, W.; Akkerman, R.; Brem, G. A critical review on recycling of end-of-life carbon fibre/glass fibre reinforced composites waste using pyrolysis towards a circular economy. *Resour. Conserv. Recycl.* **2018**, *136*, 118–129. [CrossRef]
49. Leclerc, P.; Doucet, J.; Chaouki, J. Development of a microwave thermogravimetric analyzer and its application on polystyrene microwave pyrolysis kinetics. *J. Anal. Appl. Pyrolysis* **2018**, *130*, 249–255. [CrossRef]
50. Rodrigues, G.G.M.; Paiva, J.M.F.; Carmo, J.B.D.; Botaro, V. Recycling of carbon fibers inserted in composite of DGEBA epoxy matrix by thermal degradation. *Polym. Degrad. Stab.* **2014**, *109*, 50–58. [CrossRef]

51. Kesson, D.; Foltynowicz, Z.; Jonas Christe´en, J.; Skrifvars, M. Microwave pyrolysis as a method of recycling glass fibre from used blades of wind turbines. *J. Reinf. Plast. Compos.* **2012**, *31*, 1136–1142. [CrossRef]
52. Delvere, I.; Iltina, M.; Shanbayev, M.; Abildayeva, A.; Kuzhamberdieva, S.; Blumberga, D. Evaluation of Polymer Matrix Composite Waste Recycling Methods. *Environ. Clim. Technol.* **2019**, *23*, 168–187. [CrossRef]
53. Gharde, S.; Kandasubramanian, B. Mechanothermal and chemical recycling methodologies for the Fibre Reinforced Plastic (FRP). *Environ. Technol. Innov.* **2019**, *14*, 100311. [CrossRef]
54. Lester, E.; Kingman, S.; Wong, K.H.; Rudd, C.; Pickering, S.; Hilal, N. Microwave heating as a means for carbon fibre re-covery from polymer composites: A technical feasibility study. *Mater. Res. Bull.* **2004**, *39*, 1549–1556. [CrossRef]
55. Deng, J.; Xu, L.; Zhang, L.; Peng, J.; Guo, S.; Liu, J.; Koppala, S. Recycling of Carbon Fibers from CFRP Waste by Microwave Thermolysis. *Processes* **2019**, *7*, 207. [CrossRef]
56. Fernández, Y.; Arenillas, A.; Angel, J. Microwave Heating Applied to Pyrolysis. In *Advances in Induction and Microwave Heating of Mineral and Organic Materials*; Grundas, S., Ed.; InTech: Rijeka, Croatia, 2011; pp. 724–752. ISBN 978-953-307-522-8.
57. Shamsuddin, S.-R.; Chee Ho, K.K.; Lee, K.-Y.; Hodgkinson, J.M.; Bismarck, A. Carbon Fiber: Properties, Testing, and Analysis. In *Wiley Encyclopedia of Composites*; John Wiley & Sons, Inc.: Hoboken, NJ, USA, 2012; pp. 1–17. [CrossRef]
58. Yatim, N.M.; Shamsudin, Z.; Shaaban, A.; Chafar, J.A.; Khan, M.J.H. Recovery of carbon fiber from carbon fiber reinforced polymer waste via pyrolysis. *J. Adv. Manuf. Technol.* **2020**, *14*, 37–47.
59. Zhu, W.; Xiao, H.; Wang, J.; Li, X. Effect of Different Coupling Agents on Interfacial Properties of Fibre-Reinforced Aluminum Laminates. *Materials* **2021**, *14*, 1019. [CrossRef] [PubMed]
60. Mao, L.; Shen, H.; Han, W.; Chen, L.; Li, J.; Tang, Y. Hybrid polyurethane and silane sized carbon fibre/epoxy composites with enhanced impact resistance. *Compos. Part A Appl. Sci. Manuf.* **2018**, *118*, 49–56. [CrossRef]
61. Matinlinna, J.P.; Lung, C.Y.K.; Tsoi, J.K.H. Silane adhesion mechanism in dental applications and surface treatments: A review. *Dent. Mater.* **2018**, *34*, 13–28. [CrossRef] [PubMed]
62. Li, Y.-F.; Yang, T.-H.; Kuo, C.-Y.; Tsai, Y.-K. A Study on Improving the Mechanical Performance of Carbon-Fiber-Reinforced Cement. *Materials* **2019**, *12*, 2715. [CrossRef]
63. Onischuk, A.; Panfilov, V.N. Mechanism of thermal decomposition of silanes. *Russ. Chem. Rev.* **2001**, *70*, 321–332. [CrossRef]
64. Jiang, B.; Zhao, L.; Guo, J.; Yan, X.; Ding, D.; Zhu, X.; Huang, Y.; Guo, Z. Improved thermal stability of methylsilicone resins by compositing with N-doped graphene oxide/Co_3O_4 nanoparticles. *J. Nanoparticle Res.* **2016**, *18*, 145. [CrossRef]

sustainability

MDPI

Article

A Study on the Influence of the Next Generation Colored Inorganic Geopolymer Material Paint on the Insulation Measurement of Concrete Building Shell

Yeou-Fong Li [1], Ya-Xuan Xie [1], Jin-Yuan Syu [1], Chih-Hong Huang [2,*], Hsin-Hua Tsai [2], Ta-Wui Cheng [3], Yen-Chun Chen [3] and Wei-Hao Lee [3]

1. Department of Civil Engineering, National Taipei University of Technology, Taipei 10608, Taiwan; yfli@mail.ntut.edu.tw (Y.-F.L.); a0985877979@gmail.com (Y.-X.X.); t9679010@ntut.org.tw (J.-Y.S.)
2. College of Design, National Taipei University of Technology, Taipei 10608, Taiwan; hsinhua425@gmail.com
3. Institute of Mineral Resources Engineering, National Taipei University of Technology, Taipei 10608, Taiwan; twcheng@ntut.edu.tw (T.-W.C.); t107799001@ntut.edu.tw (Y.-C.C.); glowing955146@gmail.com (W.-H.L.)

* Correspondence: huangch@mail.ntut.edu.tw

Abstract: Many studies have shown that paint with reflective heat can effectively reduce the temperature of the building envelope and reduce the future energy consumption of the building. This study inspired the next-generation inorganic geopolymer material (IGM) color paint without volatile matter, which could be applied on concrete surfaces to reduce energy consumption in warm seasons. In this study, a total of five insulating IGM paints, white, red, green, blue, and yellow, were applied to a 50 cm × 50 cm × 12 cm concrete slab top surface. The highest average light reflectance of all the paints was 87.5% of white IGM paint, which was higher than plain concrete (36.4%). The heat flux and surface temperature were examined in the laboratory, and those test results were verified outdoor. The results showed that the IGM paints could effectively reduce the surface temperature and heat flux of the upper and lower surfaces of concrete slabs, and the white colored IGM paint was the best performer among all five colors, whereas the heat storage coefficient (S_f) of red, white, yellow, blue, and green IGM painted concrete slabs were 0.57, 0.53, 3.62, 2.95, and 1.91 $W \cdot m^{-2} \cdot K^{-1}$, respectively, lower than plain concrete (24.40 $W \cdot m^{-2} \cdot K^{-1}$). This coefficient was presented to externalize the thermal admittance. The overall measurement results showed that the concrete slab with colored IGM paints had better heat insulation ability than the plain concrete slab, especially in white IGM paint.

Keywords: inorganic; geopolymer; heat storage; light reflectivity; heat flux; sustainable cities and communities

Citation: Li, Y.-F.; Xie, Y.-X.; Syu, J.-Y.; Huang, C.-H.; Tsai, H.-H.; Cheng, T.-W.; Chen, Y.-C.; Lee, W.-H. A Study on the Influence of the Next Generation Colored Inorganic Geopolymer Material Paint on the Insulation Measurement of Concrete Building Shell. *Sustainability* **2022**, *14*, 164. https://doi.org/10.3390/su14010164

Academic Editor: Asterios Bakolas

Received: 24 November 2021
Accepted: 21 December 2021
Published: 24 December 2021

1. Introduction

The problem of global warming caused by greenhouse gas emissions has continued to worsen in the last few decades. In hot seasons, the demand for air conditioners is increasing. Such demand is boosting the human consumption of energy, with carbon emissions swelling, which has intensified global warming. Without effective reduction of greenhouse gas emissions, the Earth will face unimaginable consequences. Thus, improving energy efficiency and reducing energy dependency are crucial and can be accomplished by effectively managing the use of energy or through insulation. The thermal capacity of building concrete envelope emerged and became one of the most important factors for urban energy balance, which could extensively affect the Urban Heat Island (UHI) in the urban climate [1,2]. It is also a key element to obtain sustainable cities and communities.

In addition to the traditional low thermal conductivity insulating bricks, the materials used for the energy-saving and heat-insulating improvement of the building shells are gradually introduced into the paints with thermal insulation effects. With light-reflective paints' effective insulation technology, many studies have shown that the light reflectance of cold-colored paint formulations was significantly higher than general paints, and the surface temperature was lower than general paints as well. The heat flux (or the thermal flux, sometimes referred to as heat flux density) was relatively low once the light-reflective paint was applied, and the surface temperature of the building wall and roof can be effectively reduced. The surface temperature of the walls was decreased enough to have an energy-saving effect [3–10]. In the past, studies on different paint colors also showed that the brighter the surface, the higher the light reflectance value created, which also significantly lowered the temperature [11–13].

After the finding, buildings in many countries were beginning to be painted with bright colored paints such as Menton in France, Newfoundland in Canada, and Bristol in the UK. The development and research of paints with both reflective heat insulation and color diversity was booming. Further findings showed that not only could the shades of the paint color affect the reflection of the sunlight, when phase change materials were added to the paints under constant temperature conditions, they could also change the state of matter and provide latent heat to achieve thermal insulation [14]. Some researchers studied the paints with reflection and heat insulation by applying them to a wide variety of buildings and roads [15–18]. There were also studies on the use of white cement paint, high-gloss reflective gray or white paints, and environmental afforestation on the building rooftops, all aimed to accelerate the light-reflecting of the sunlight. In comparison, white paint proved to be more effective in reducing temperature than other heat-insulating methods [19–21]. Additional studies had pointed out that dust deposited on the roof surface was another effective method, which accelerated thermal conductivity and increased building cooling energy consumption [22].

The building envelope was one of the main thermal regenerators of the building; therefore, the quality of the heat storage capacity of the casing could also significantly improve any discomfort that is caused by the thermal in the room [23]. It has also been suggested from the past studies that the heat storage capacity affects the surface colors of different types of shells and the absorption, release, and temperature fluctuations of the shells facing them [24]. Working as the first line of defense, the outer casing was responsible for reducing the energy consumption of buildings. It was introduced with low heat storage technology, in addition to reflecting and blocking the heat of outdoor visible and infrared light that was passing through the building shell to the laboratory to reduce heat storage. Meanwhile, it could also embellish the city and improve energy savings.

Considering that available commercial paints were mostly composed of organic polymer resins, which allow them to be affected by ultraviolet radiation under long-term sunlight, resulting in aging, deterioration, and reduced thermal insulation properties. In the EU and the United States, the content of volatile organic compounds (VOCs) that might cause disease in paints must be strictly limited to below 30 g/L and 50 g/L, respectively. To meet the restrictions, the thermal insulation paints used in this study did not contain any volatile VOCs. The IGMs used in this experiment were consisted of compositions such as whetstone, kaolin, calcium magnesium carbonate, titanium dioxide and natural minerals, and an alkali solution. The alkali solution was made of glass, alkaline metal salt, nano-cerium dioxide, etc. It had a heat-reflective insulation capability and could be applied to concrete building shells as insulation materials and meet the energy-saving requirements of concrete building shells.

The process of the experiment was to first measure the reflection and heat insulation function of the IGM paints, while also considering a color that was suitable for the building. By adding 6% inorganic pigment blend, a variety of colors were produced for options to maintain the aesthetic appearance of the building shell. No volatile substances (VOCs) were emitted during construction and procedures. At the same time, the nano-mineral powder was added to overcome the problem when the porous surface was susceptible to dirt and affects the heat insulation. In all, this study conducted a series of material performance tests on white, red, green, blue, and yellow IGM paints in both laboratory and outdoor measurements to investigate the thermal insulation properties of this IGM paint.

2. Experimental Measurement and Theory of Thermal Property

2.1. Light Reflectance Measurement

The light reflectivity possessed an important effect on the reduction of building energy consumption and surface temperature reduction, which was measured with the UV/VIS/NIR Spectrometer (LAMBDA 900, PerkinElmer Inc., Waltham, MA, USA) based on ASTM E1331-15, as shown in Figure 1. The spectrometer was used to measure the reflectivity of five colored IGM paint surfaces with the paint on the top of the steel plate (50 mm × 50 mm) with a thickness of 0.3 mm. The light reflectivity measurements of all five colored IGM paint surfaces and a plain concrete surface were repeated three times. The light source was separated by a diffraction grating and then irradiated into the paint sample through the instrument, and the light reflectance was measured according to the magnitude of the current. It could accurately measure visible light (VL, wavelength 390–700 nm), near-infrared radiation (NIR, wavelength 700–2000 nm), and the accuracy of reflective light grating in spectral ranges was 1 nm. The measurement underwent qualitative analysis and relatively reflectance relative to the preset values of the spectrometer in this article. According to the reference, the hemispherical solar irradiance on a 37° tilted surface ($W \cdot m^{-2} \cdot nm^{-1}$) between 390 nm to 2000 nm was 91.60% of 280 nm to 4000 nm [25].

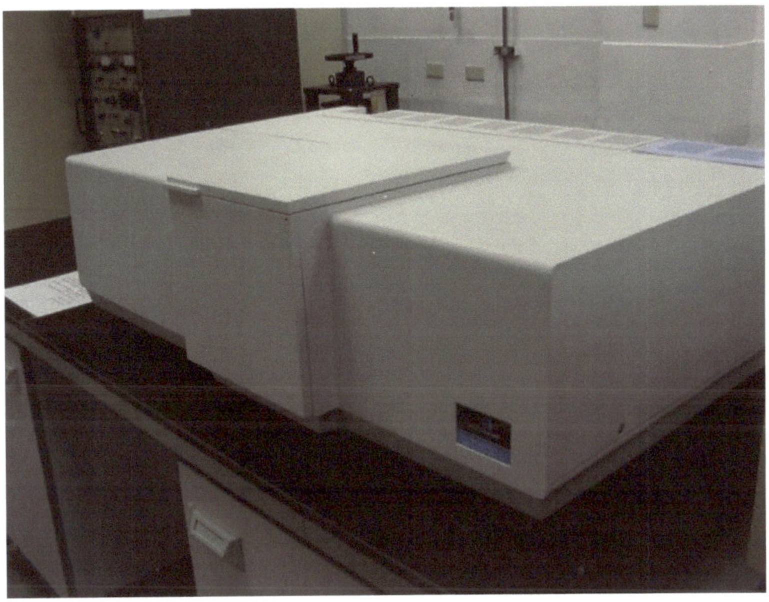

Figure 1. UV/VIS/NIR Spectrometers.

2.2. Laboratory Measurement of Heat Flux and Temperature

A total of seven 50 × 50 × 12 cm concrete slabs were prepared for the laboratory's thermal insulation performance measurements; each concrete slab was painted with two layers of IGM paints with a total thickness of 0.3 mm. The density of the concrete slabs was 2592 kg/m^3. The seven test concrete slabs included one plain concrete and six others each painted in white, red, green, blue, and yellow IGM paint, and commercial white organic paint (white paint (com'l)). The IGM powder was a rutile type of titanium dioxide (TiO_2) component with a heat-insulating reflection effect, and it was permeable, which allowed it to produce various colors by adding different metal oxide pigments. The test used an IGM powder mix which contained 6% colorant and alkali liquid, which included water, silicate solution, alkali salt, and silicon dioxide (SiO_2). The weight ratio of the mixed powder to liquid was 1:1, and it was uniformly mixed to form a heat-insulating material for paint.

Because solar irradiance contains 44% visible light and 53% infrared radiation, three 500 W halogen lamps and one 600 W infrared lamp were used to simulate the total irradiance of VL and NIR that emit to the concrete buildings. The heights of the lighting devices were erected to 20 cm for the halogen lamps and to 35 cm for the infrared lamp, respectively. The irradiance of the visible light and infrared radiation were 609 W/m^2 and 775 W/m^2, respectively; the overall irradiance was 1384 W/m^2 in this experiment measurement. As a reference, in the Taipei area, the highest solar irradiance ever recorded in the recent ten years inferred to the Central Weather Bureau in Taiwan was 1409.7 W/m^2 at 1 p.m. on 10 July 2020. The heat flow patch (PHFS-01e, FluxTeq, Blacksburg, VA, USA) and the thermocouple (K-type, model TPK-01BN, TECPEL, New Taipei City, Taiwan) were attached to both the front and back sides of the concrete slab, and the heat flux and surface temperature of the upper and lower portions were measured. The schematic diagram of the heat flux, lamps, and surface temperature measurement and the test setup photo are shown in Figure 2a,b, respectively. Thermal insulation performance measurements of the seven concrete slabs were repeated three times in the laboratory.

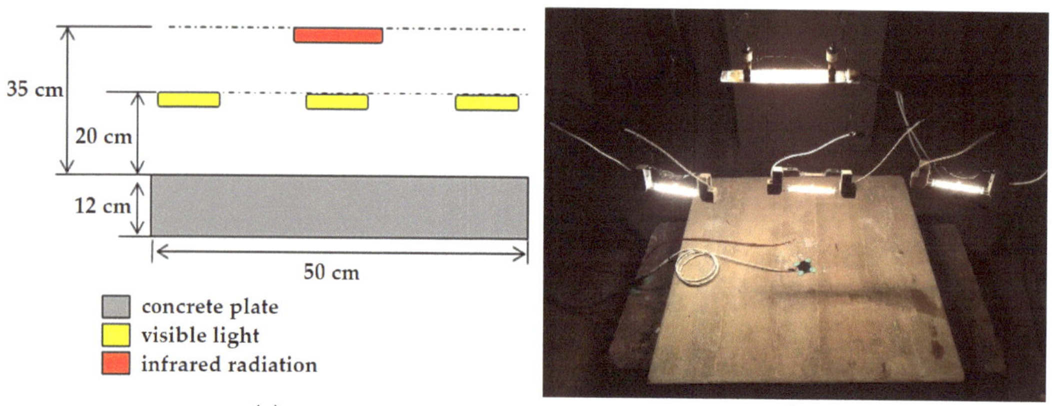

Figure 2. The heat flux and surface temperature measurement of concrete slab in the laboratory: (**a**) Schematic diagram; (**b**) Test setup photo.

2.3. Outdoor Measurement of Heat Flux and Temperature

The seven concrete slabs were placed on the rooftop of the Civil Engineering building at the National Taipei University of Technology to simulate an outdoor condition with the ambient temperature at about 30 °C, then, the heat flux and temperature on the upper and lower surface of the painted concrete and the plain concrete specimens were measured and recorded throughout the experiment. In the meantime, the ambient temperature, wind speed, and humidity of the weather were recorded, and the solar irradiance was measured by a pyranometer. After comparing the five colored IGM paint concrete slabs with the commercial white paint and the plain concrete slabs, the thermal insulation performance results were obtained within a 24-h time frame in this outdoor experiment. Figure 3 demonstrates the outdoor setup of the concrete slabs including the pyranometer, heat flux sensor, and surface thermograph.

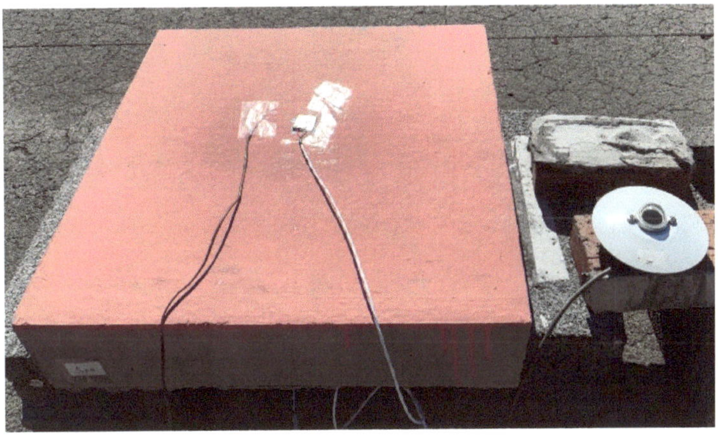

Figure 3. Outdoor concrete slab measurement setup.

2.4. Heat Absorption, Conduction, and Storage

Figure 4 showed the upper and lower heat fluxes and temperatures, heat transfer, and heat storage behavior of the concrete slab. In this study, solar irradiance was a measurement value of the solar radiation after passing through the atmosphere to the concrete surface. The upper surface of the concrete slab came in contact with the solar irradiance; part of irradiance was first reflected by the surface material, and the rest of the radiance was then converted to thermal energy and absorbed by the concrete slabs. The heat flux and temperature were two common physical behaviors of heat absorption.

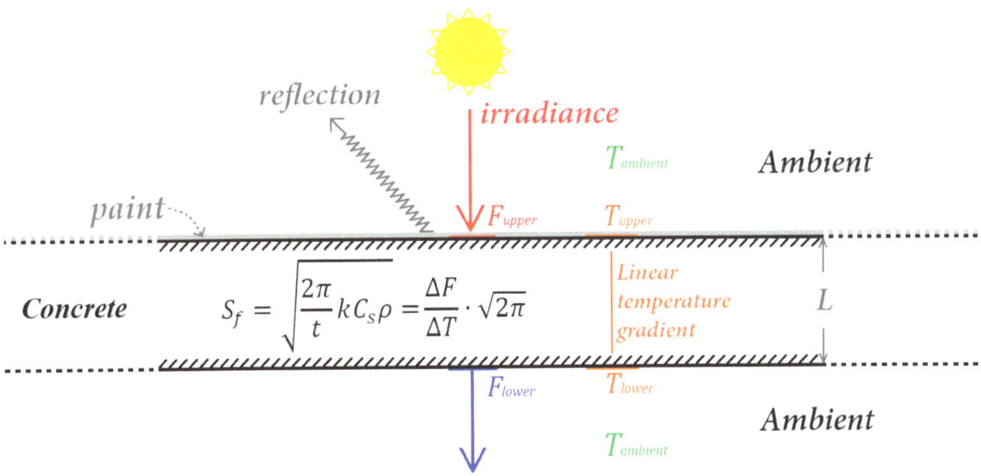

Figure 4. The heat flux, transfer, and storage behavior of concrete slab.

According to thermodynamics, the thermal capacity was defined by the quantity of heat or energy a subject needed to increase a unit temperature change. It was related to the specific heat capacity and mass of the subject as in Equation (1). According to some research [26,27], the heat (Q) change in the system was related to the temperature change, subject mass, and its specific heat capacity. The quantity of heat in absorption and conduction and the specific heat capacity of the subject were hard to measure; thus,

Fourier's law of interface heat transfer was used to deduce the heat flux, reception surface area, and reception time length are shown in integral form of Equation (2):

$$C = \frac{Q}{\Delta T} = m \cdot C_s \tag{1}$$

$$Q = m \int_{T_0}^{T_x} C_s(T) \cdot dT = A \int_{t_0}^{t_x} F(t) \cdot dt \tag{2}$$

where C is heat capacity ($J \cdot K^{-1}$); Q represents the heat or energy added or subtracted into the system (J); ΔT is temperature changes (K); m is the mass of the concrete slab (kg); and C_s is specific heat capacity ($J \cdot kg^{-1} \cdot K^{-1}$). In Equation (2), T_0 and T_x are the initial and final temperatures (K); $F(t)$ is the heat flux ($W \cdot m^{-2}$); t_0 and t_x are the initial and final time (s), respectively, and A is the area between concrete and ambient air boundary (m^2).

From Equations (1) and (2), the specific heat capacity C_s of the subject could be rebuilt based on the heat flux, as shown in Equation (3):

$$C_s = \frac{F(t) \cdot A}{m \cdot \frac{dT}{dt}} = \frac{\Delta F \cdot A \cdot t}{m \cdot \Delta T} \tag{3}$$

Equation (4) defined the relationship between the heat flux, the thermal conductivity coefficient, and the temperature gradient as in Fourier's law in one dimension [28–30]. Regarding Equation (4), the thermal conductivity coefficient can be calculated by heat flux, thickness, and temperature variation of concrete specimens, as shown in Equation (5):

$$F = -k \cdot \nabla T = k \cdot \frac{-\Delta T}{L} \tag{4}$$

$$k = \frac{F \cdot L}{-\Delta T} \tag{5}$$

where L is the thickness of concrete slab (m), and k is the thermal conductivity coefficient.

Moreover, the heat storage coefficient that had been studied in the past was used here to clarify the heat storage ability in heat transfer processes as in Equation (6) [31] with unit $W \cdot m^{-2} \cdot K^{-1} \cdot s^{0.5}$ and Equation (7) [32] with unit $W \cdot m^{-2} \cdot K^{-1}$:

$$S = \sqrt{k C_s \rho} \tag{6}$$

$$S_f = \sqrt{\frac{2\pi}{t} k C_s \rho} \tag{7}$$

where S is heat storage coefficient ($W \cdot m^{-2} \cdot K^{-1} \cdot s^{-0.5}$); S_f is the quantity of heat storage coefficient effect of periodic fluctuation ($W \cdot m^{-2} \cdot K^{-1}$).

It was assumed that the thermal conductivity of the concrete slab was constant, and the heat transfer within the concrete was linear. Under such assumption, the heat storage coefficient of the tested concrete slabs for 20 h in the laboratory was defined as S and S_f as in Equations (10) and (11), which described the heat absorption and its release based on the heat flux and temperature differences (Equations (8) and (9)). As a result, the heat storage coefficient S_f of painted and plain concrete slabs was presented:

$$\Delta F = F_{upper} - F_{lower} \tag{8}$$

$$\Delta T = T_{mean} - T_{ambient} = \frac{T_{upper} + T_{lower}}{2} - T_{ambient} \tag{9}$$

$$S = \sqrt{k C_s \rho} = \sqrt{\frac{\Delta F \cdot A \cdot t}{m \cdot \Delta T} \cdot \frac{\Delta F \cdot L}{\Delta T} \cdot \rho} = \frac{\Delta F}{\Delta T} \cdot \sqrt{t} \tag{10}$$

$$S_f = \sqrt{\frac{2\pi}{t} k C_s \rho} = \sqrt{\frac{2\pi}{t} \cdot \frac{\Delta F \cdot A \cdot t}{m \cdot \Delta T} \cdot \frac{\Delta F \cdot L}{\Delta T} \cdot \rho} = \frac{\Delta F}{\Delta T} \cdot \sqrt{2\pi} \quad (11)$$

where F_{upper} is the heat flux from paint to concrete slab under the light ($W{\cdot}m^{-2} = J{\cdot}m^{-2}{\cdot}s^{-1}$); F_{lower} is the heat flux from the back of concrete to the ambient ($W{\cdot}m^{-2} = J{\cdot}m^{-2}{\cdot}s^{-1}$); T_{mean} is the mean temperature in concrete under a linear gradient that changed over time (K); $T_{ambient}$ is the ambient temperature that changed over time (K).

3. Light Reflectance Measurement Results

From the studies, the light reflectance of commercial organic white paints could reach 80 to 86% [4,12]. The rest, including the light reflectance measurement results of the five colored IGM painted surfaces and the plain concrete, were shown in Figure 5 and Table 1. As shown in Figure 5 and Table 1, the first finding indicated that between the SR (390–2000 nm) spectral range, the average light reflectance of the white IGM painted surface reached 87.5%, followed by yellow, blue, green, and red IGM paints, and the plain concrete was 36.4%. The second finding indicated that IGM was tested higher in the light reflectance values of NIR than in VL.

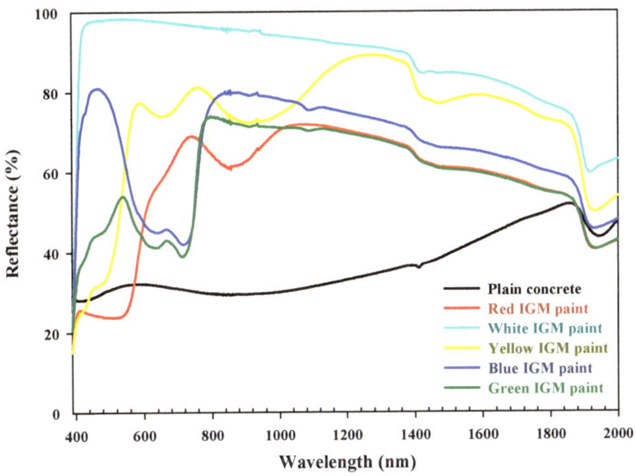

Figure 5. The light reflectance distribution of plain concrete and colored IGM paints in the wavelength range of 390–2000 nm.

Table 1. Light reflectance of the plain concrete and colored IGM paints under each light band.

Light	Wavelength	Average Light Reflectance (%)					
		Plain Concrete	White	Yellow	Blue	Green	Red
VL	390–700 nm	30.7	94.0	53.9	59.6	43.7	36.6
NIR	700–2000 nm	37.7	86.0	76.6	67.1	61.9	62.0
SR	390–2000 nm	36.4	87.5	72.2	65.6	58.4	57.1

4. Laboratory Heat Flux and Surface Temperature Measurements

4.1. Laboratory Heat Flux Measurement Results

For 20 h, the upper surface of all seven concrete specimens were shined under both halogen lamp (visible light, VL) and near-infrared lamp (NIR) simultaneously. Figure 6 showed the heat flux changes in both the upper surface and lower surface of the seven concrete slab specimens over time. Table 2 showed the heat flux changes in both the upper surface and lower surface of the seven concrete slab specimens in the 20th hour. In short, IGM painted specimens showed lower heat flux in both charts.

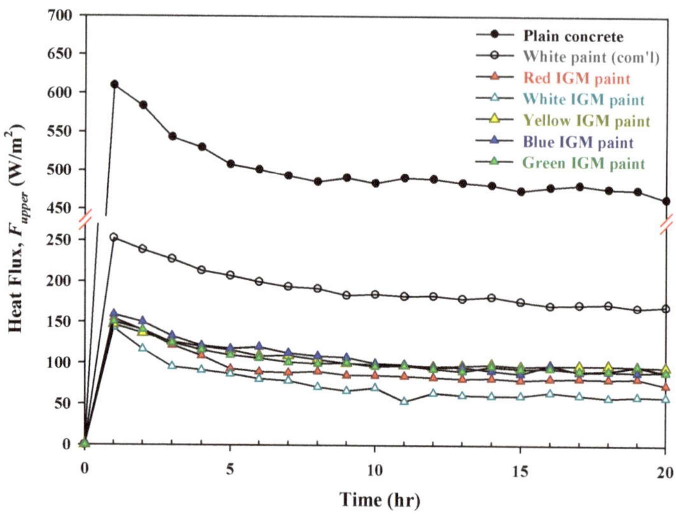

(**a**)

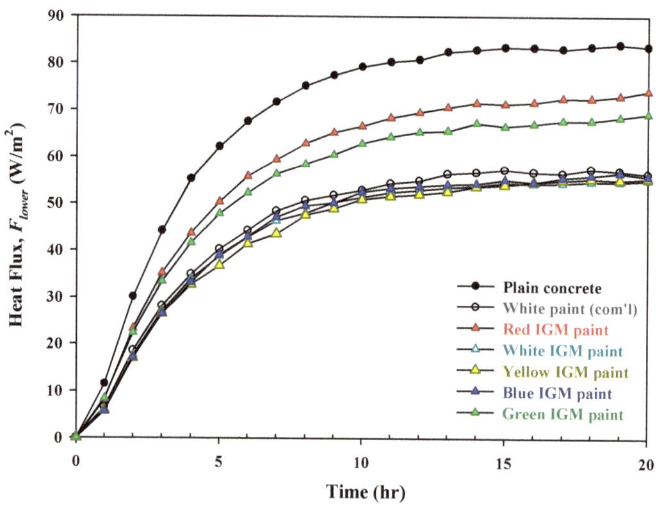

(**b**)

Figure 6. The heat flux of concrete specimens under SR over time: (**a**) Upper surface heat flux; (**b**) Lower surface heat flux.

Table 2. The upper surface and lower surface heat fluxes in the 20th hour.

Color	Upper Surface Heat Flux (W/m^2)	Lower Surface Heat Flux (W/m^2)
Plain Concrete (unpainted)	463.40	83.50
White Paint (com'l)	170.10	56.35
Red IGM Paint	73.99	74.06
White IGM Paint	59.07	55.04
Yellow IGM Paint	95.40	55.40
Blue IGM Paint	89.88	55.83
Green IGM Paint	89.70	69.23

In detail, the results of the heat fluxes of the specimens stated that the IGM paint was more effective in blocking the heat created by the halogen lamp and the infrared lamp compared to the commercial white paint and to the plain concrete in the upper surface. The lower surfaces showed similar results regarding the IGM paint, the commercial white paint, and then the plain concrete. Note that the lower surfaces of the seven specimens were not painted. In other words, the results proved that the IGM paint had a better heat barrier property that could effectively reflect the radiant of the portion it covered and slowed the heat flow into the lower surface of the concrete slabs as well.

4.2. Laboratory Surface Temperature Measurement Results

Figure 7 showed the duration of the temperature changes of both the upper and the lower surface of concrete specimens under the illumination of a halogen lamp and infrared lamp in the laboratory, and Table 3 showed the measured values of the steady-state temperature of the upper and lower surfaces under the same settings.

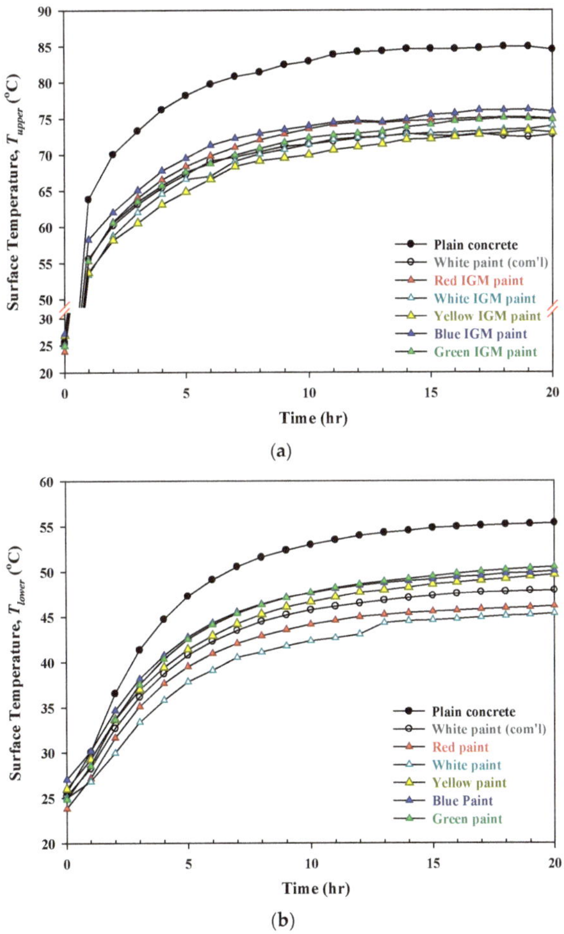

Figure 7. The surface temperature of concrete specimens under VL and NIR over time: (**a**) Upper surface temperature; (**b**) Lower surface temperature.

Table 3. The upper surface and lower surface temperatures of the concrete slabs under VL and NIR at a steady-state condition.

Color	Temperature (°C)		
	Average Ambient	Upper Surface	Lower Surface
Plain Concrete (unpainted)	29.77	84.59	54.82
White Paint (com'l)	30.49	72.60	47.39
Red IGM Paint	28.57	74.83	45.65
White IGM Paint	31.18	73.10	44.66
Yellow IGM Paint	32.04	72.38	48.75
Blue IGM Paint	32.21	75.53	49.33
Green IGM Paint	30.14	74.28	49.70

All seven concrete specimens were irradiated under the halogen lamp and the infrared lamp, and the temperature of each concrete slab specimen's upper surface was measured and shown in Figure 7a and Table 3. The lowest temperature measured was the concrete slab painted with the yellow IGM paint, followed by the ascending order of white IGM, white paint (com'l), red IGM, green IGM, and blue IGM plain concrete specimens. In comparison, the lower surface temperature of each concrete slab specimen was also measured, and the lowest temperature was the white painted concrete slab specimen, followed by the red IGM, white paint (com'l), yellow IGM, blue IGM, green IGM, and then the plain concrete specimens in an ascending order in Figure 7b and Table 3.

In the concrete slabs under the illumination of the light source experiment, the white IGM paint had the best thermal insulation ability of all the colored IGM paints, and it was superior to the commercially available white paint and plain concrete. In general, the thermal insulation abilities of the painted concrete specimens were all better than the plain concrete.

4.3. *Laboratory Surface Temperature Measurement Results after Turn off the Light*

Figure 8 showed the temperature drop gradation of the specimens to describe the heat release of paint concrete slabs when the lamps were turned off. Each slope showed the gradation in temperature changes for each concrete specimen from their starting temperature (after 20 h of the illumination) until 30 min later.

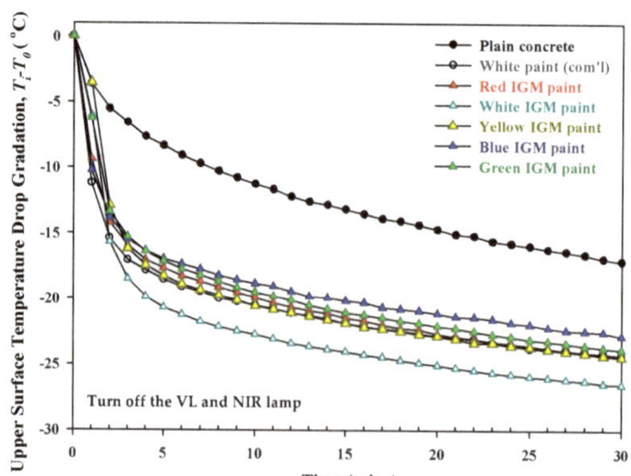

(a)

Figure 8. *Cont.*

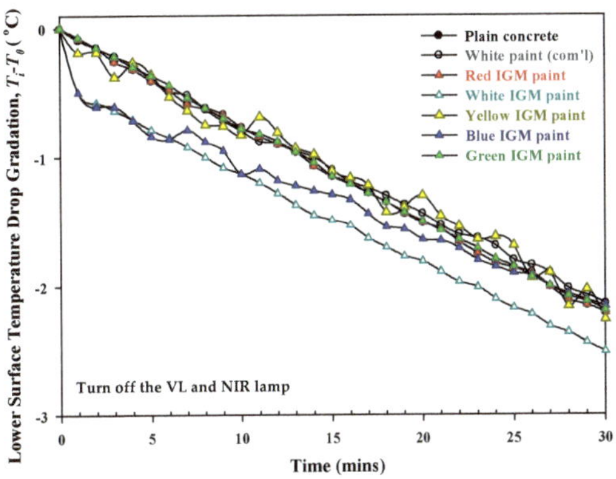

(**b**)

Figure 8. The surface temperature drop gradation of the concrete slabs over time: (**a**) Upper surface; (**b**) Lower surface.

According to Figure 8, the upper surface temperature of the plain concrete was the worst among all specimens in cooling off, whereas the upper surface cooling rate of the IGM paints were over 1.72 °C·min^{-1} better than the plain concrete in the first 5 min of the 30 min cooling process. In addition to that, the heat release of the white IGM paint concrete slab was better than the commercial white paint concrete slab. Regarding the lower surface measurements, the cooling outcomes of the white and blue colors were the best. It showed that the painted concrete slabs could cool off more rapidly than the plain concrete slab. That is to say that the paint layer could effectively dissipate the heat stored inside the specimen.

4.4. Thermal Admittance Calculation and t-Test Analysis

Thermal admittance is an important index in describing the heat absorbing and releasing in relation to spaces of the building materials over time. In this article, the changes of the heat storage coefficient were used to label the thermal admittance. Table 4 showed the specific heat capacity and heat storage coefficient of the colored IGM paint specimens and the plain concrete specimen based on the laboratory test results. The plain concrete had the highest heat storage coefficient (Thermal Admittance), while the white IGM paint concrete specimen had the lowest. Evidently, IGM paint could reduce the quantity of the heat absorption of concrete. As the heat flux into a concrete specimen decreased, the heat storage coefficient decreased as well, and such an interaction is shown in Figure 9. The heat absorption accumulation increased as time continued; however, the decrease in the unit heat absorption quantity was found in the experiment as shown in Figure 10, which agreed with the concept of when the heat storage of a building concrete gradually decreased and the heat flux slowed under SR over time.

Table 4. Calculation of specific heat capacity and heat storage coefficient at a steady-state condition (last 5 h average) under the light.

Color	T_{mean} (°C)	$T_{ambient}$ (°C)	ΔT (°C)	ΔF (W·m^{-2})	C_s (J·kg^{-1}·K^{-1})	S (W·m^{-2}·K^{-1}·s$^{0.5}$)	S_f (W·m^{-2}·K^{-1})
Plain Concrete (unpainted)	69.94	29.74	40.203	391.34	126.96	9.73	24.40
White Paint (com'l)	60.19	30.66	29.533	114.18	50.43	3.87	9.69
Red IGM Paint	60.49	28.70	31.785	7.21	2.96	0.23	0.57
White IGM Paint	59.26	31.42	27.841	5.88	2.76	0.21	0.53
Yellow IGM Paint	61.09	32.28	28.806	41.66	18.86	1.45	3.62
Blue IGM Paint	62.88	32.50	30.382	35.73	15.34	1.18	2.95
Green IGM Paint	62.54	30.44	32.096	24.41	9.92	0.76	1.91

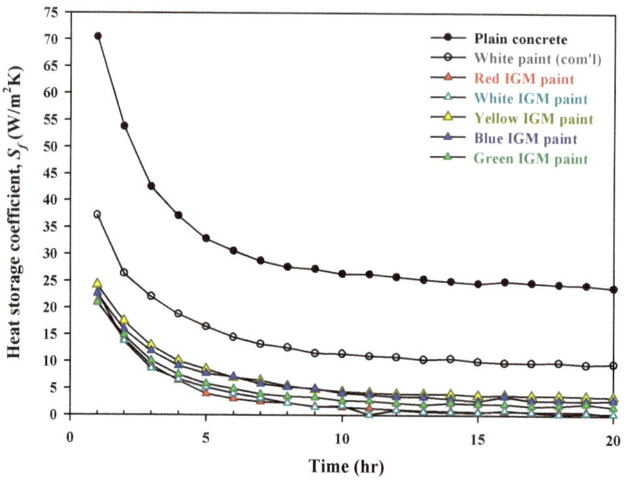

Figure 9. The heat storage coefficient of plain concrete slab and six painted slabs over time.

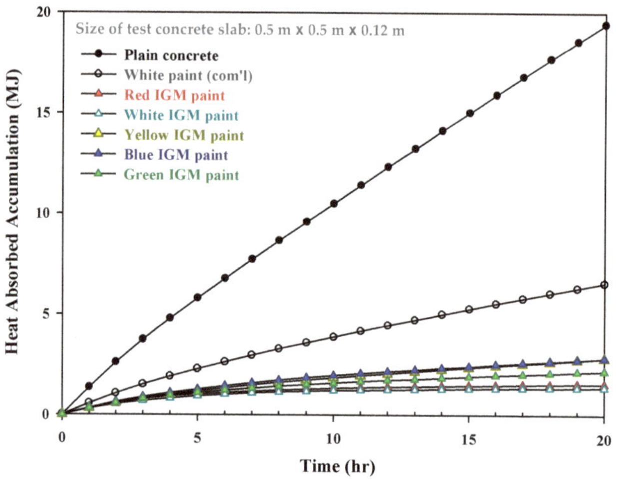

Figure 10. The heat absorbed accumulation of plain concrete slab and six painted slabs over time.

The cooling rate of the plain concrete slab was slower than the other painted concrete slabs. After the painted concrete slabs were illuminated by the light, some of the heat could be reflected first due to the influence of the paint layer. So, the heat absorbed by the test body was relatively reduced. In consequence of the lower amount of heat travelling to the lower surface, the temperature was lower in respect to the effect of the paint layer. This proved that the next generation of IGM paint has good reflective heat insulation and low heat storage capacity. If it could be effectively applied to the insulation of building shells, it was believed that it would be able to improve the energy efficiency of urban buildings and reduce the energy consumption of all buildings.

According to the above experimental results above, the t-test analysis between the VL reflectance, SR reflectance, and heat storage coefficient of different-colored IGM paints and plain concrete were calculated in this study. The Pearson correlation coefficient (Pearson's R) between the reflectance of the SR, VL, NIR and heat storage coefficient (S_f) were about -0.763, -0.504, and -0.819, respectively; the t-values were about -1.313, -1.340, and -1.306, respectively. In addition, the p-value (one-tailed) were about 0.123, 0.119, and

0.124, respectively, and all the *p*-value were higher than 0.05. These results designated that the reflectance of the color paints might not be a primary effective factor of the thermal absorption of concrete. However, as stated by the results of the correlation coefficient, the light reflectance of color paints had the negative relationship with the heat storage coefficient that was crucial enough to affect the outcomes. According to physics, light is a major form of energy radiation; therefore, increasing the light reflectance was a good way to reduce the heat absorption of concrete. The results also indicate that IGM paints show a promising interaction between the NIR reflectance and the heat storage coefficient (S_f) since heat radiation was often presented as a type of infrared radiation. Namely, the NIR reflectance and the heat storage coefficient should be carefully considered and analyzed in thermal absorption obstruction assessments for painted concrete.

5. Outdoor Heat Flux and Surface Temperature Measurements

All concrete slabs were placed outdoors for 24 h to measure heat flux and temperature properties, and the results were presented in subsequent sections.

5.1. Outdoor Heat Flux Measurement Results

The heat fluxes of the upper and lower surfaces are shown in Figure 11. In addition, the solar irradiance is shown in Figure 12. Table 5 shows the maximum values of the upper surface and lower surface heat fluxes of concrete slabs over 24 h. The results indicate that the upper surface and lower surface heat fluxes of the painted specimens were lower than that of the plain concrete, even when the solar irradiance in the colored paint specimens was higher than in the plain concrete specimens according to the pyranometer. The white IGM paint had the best heat insulation performance among all specimens, while the plain concrete specimen had the poorest thermal insulation. The outdoor measurement results of all concrete specimens were similar to the laboratory measurement results in comparison.

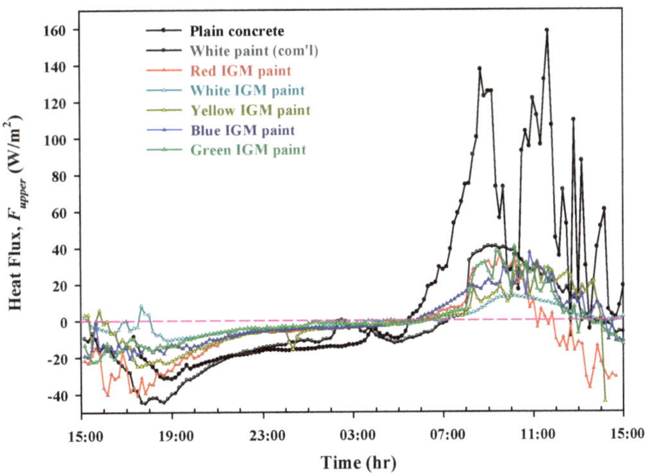

(a)

Figure 11. *Cont.*

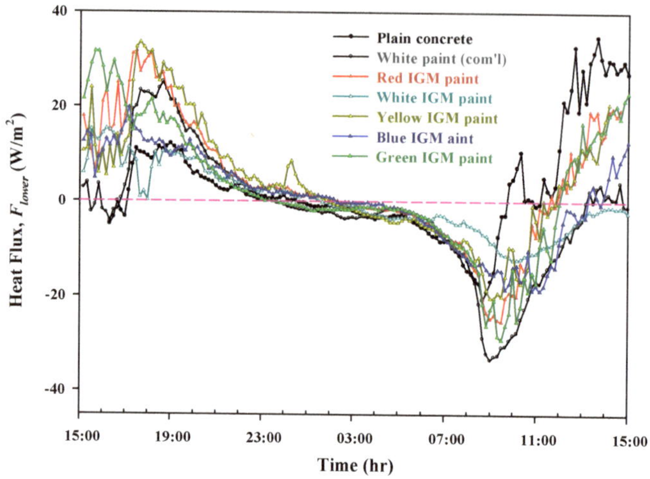

(**b**)

Figure 11. The heat flux of the concrete slabs during 24 h outdoor measurement: (**a**) Upper surface; (**b**) Lower surface.

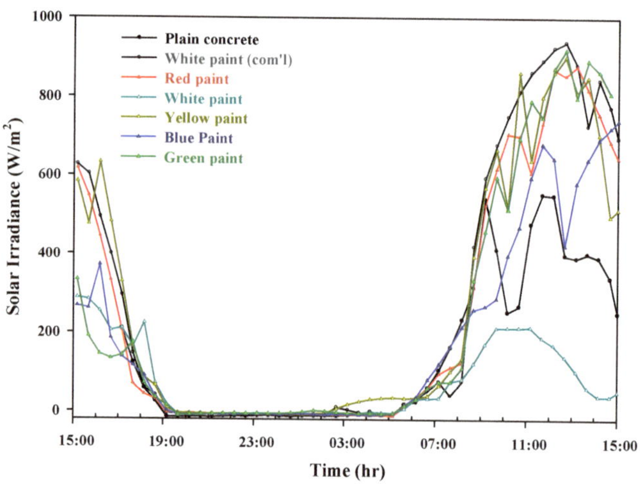

Figure 12. The solar irradiance of the concrete slabs during 24 h outdoor measurement.

Table 5. The maximum values of the upper surface and lower surface heat fluxes of concrete slabs during the 24 h of observation.

Color	Upper Surface Heat Flux (W/m^2)	Lower Surface Heat Flux (W/m^2)
Plain Concrete (unpainted)	158.10	34.80
White Paint (com'l)	40.58	25.16
Red IGM Paint	35.96	31.66
White IGM Paint	13.49	15.28
Yellow IGM Paint	32.00	33.53
Blue IGM Paint	37.17	19.72
Green IGM Paint	40.47	31.60

5.2. Outdoor Surface Temperature Measurement Results

Sometimes, the sun was obscured by the clouds during the daytime, which would affect the solar irradiance and temperatures. The average temperature of the upper and lower surfaces and the solar irradiance of the tested specimens were compared. The results were divided into two time sections, one from 10:00 to 17:00 (daytime) and another from 19:00 to 5:00 (nighttime).

Table 6 shows the average temperature and temperature changes of the upper surfaces of all seven specimens during the daytime. The white IGM paint had the lowest average temperature and the smallest changes of temperature, followed by the blue IGM, the green IGM, the commercial white, the red IGM, the yellow IGM, and the plain concrete had the highest average temperatures and the biggest temperature changes. Similarly, as seen in Table 6, the lower surface temperature differences of the six painted concrete slab specimens were all lower than the plain concrete. In Table 7, the nighttime average temperature of both the upper and the lower surfaces of the six painted concrete specimens were all lower than the plain concrete specimen. Moreover, the average temperature and temperature difference of the white IGM paint specimen had the lowest among the six painted specimens. Based on the data of the solar irradiance recorded in the experiment, the white IGM paint specimen had a lower solar irradiance than all the others due to the weather conditions.

Table 6. Average daytime (10:00–17:00) temperature, humidity, and wind speed mean values of the outdoor measurement.

Color	Ambient Temperature (°C)	Humidity (%)	Wind Speed (m/s)	Upper Surface Temperature (°C)	Lower Surface Temperature (°C)
Plain Concrete (unpainted)	31.12	56.94	1.43	42.24	40.90
White Paint (com'l)	34.17	48.47	4.30	42.42	40.07
Red IGM Paint	35.80	43.84	2.38	45.97	41.91
White IGM Paint	33.02	57.47	2.46	35.41	35.23
Yellow IGM Paint	33.89	55.29	4.23	44.92	39.88
Blue IGM Paint	34.44	50.67	3.02	40.75	38.16
Green IGM Paint	33.70	56.50	3.23	42.37	38.75

Table 7. Average nighttime (19:00–05:00) temperature, humidity, and wind speed mean values of the outdoor measurement.

Color	Ambient Temperature (°C)	Humidity (%)	Wind Speed (m/s)	Upper Surface Temperature (°C)	Lower Surface Temperature (°C)
Plain Concrete (unpainted)	26.49	81.72	1.16	31.64	33.98
White Paint (com'l)	30.06	68.90	2.67	29.47	30.46
Red IGM Paint	29.38	72.80	1.28	29.19	30.32
White IGM Paint	31.02	68.46	1.69	30.63	31.49
Yellow IGM Paint	29.45	82.07	3.50	30.36	31.17
Blue IGM Paint	29.32	78.75	1.50	29.82	30.86
Green IGM Paint	29.67	74.30	1.95	28.84	29.65

In the outdoor measurement, the white IGM paint specimen had the best thermal insulation ability in all colors. The thermal insulation ability of the plain concrete had the poorest insulation ability among all specimens. These outdoor measurement results were similar to the laboratory measurement results. Regardless of either the laboratory or the outdoor measurements, the thermal insulation ability of the IGM-painted concretes were all better than the plain concrete and commercial paint concrete.

6. Conclusions

Solar radiation is the main heat source for most concrete building rooftops; therefore, finding out how to minimize the concrete's heat absorption and ameliorate the heat releasing of the concrete buildings has become crucial in order to achieve a greener building code. This study attempted to improve the light reflectivity of concrete surfaces with color IGM paints to reduce the heat absorption of concrete. Some conclusions of this study are stated as follows:

1. The average light reflectance values of various color paint materials were obtained by the light reflectance test. In the range of 390–2000 nm, the highest average light reflectance value of all the covering materials was the white IGM at 87.5%, and the light reflectance of the plain concrete was 36.4%.
2. It is known from the thermal insulation results in the laboratory that the IGM paint had good thermal insulation properties. Under the influence of environmental factors, the measurement results were inevitably challenged. Nevertheless, the results of the experiment were still considered to be objective under the circumstances. That is, under the irradiance of simulated sunlight, the thermal insulation performance of the concrete specimens painted with the IGM paints were better than the plain concrete specimen.
3. Thermal admittance was described in place of the changes of the heat storage coefficient in this research. The results indicated that the IGM paints could reduce the quantity of heat absorption of the concrete slabs. The heat storage coefficient of red, white, yellow, blue, and green IGM-painted concrete slabs were 0.57, 0.53, 3.62, 2.95, and 1.91 $W{\cdot}m^{-2}{\cdot}K^{-1}$, respectively, which were lower than the plain concrete specimen (24.40 $W{\cdot}m^{-2}{\cdot}K^{-1}$). Based on the changes of the heat storage coefficient over time, it was confirmed that the heat absorption of specimens through the IGM paints and then into the concrete under the light was slowed, and it was lower than the commercial white paint specimen and the plain concrete specimen.
4. The results of outdoor measurement showed that the IGM painted specimens had good thermal insulation ability, and these results were similar to the measurement results in the laboratory. The concrete specimens painted with IGM had better thermal insulation performance than the plain concrete specimen. Furthermore, the thermal insulation performance of the white IGM paint was also better than the commercial white paint.
5. Through various physical properties and thermal insulation measurement results, it was clear that the next generation of colored IGM paints had high reflectivity, good thermal insulation, and low heat storage capacity. The next generation of colored IGM paint could be applied to various building shells and especially to concrete building shells in sustainable cities and communities. Its strong and practical characteristics could effectively reduce the indoor temperature and attain an energy-saving effect in subtropics.

Author Contributions: Conceptualization, Y.-F.L.; data curation, Y.-X.X. and Y.-C.C.; formal analysis, Y.-X.X., J.-Y.S. and H.-H.T.; investigation, Y.-X.X., J.-Y.S. and W.-H.L.; methodology, Y.-F.L. and C.-H.H.; project administration, Y.-F.L. and C.-H.H.; supervision, Y.-F.L., W.-H.L. and T.-W.C.; writing—original draft, Y.-X.X. and T.-W.C.; writing—review and editing, Y.-F.L. and C.-H.H. All authors have read and agreed to the published version of the manuscript.

Funding: This research was funded by the Ministry of Education of Taiwan, under contract No. L7081101-4, and the "Research Center of Energy Conservation for New Generation of Residential, Commercial, and Industrial Sectors" from the Ministry of Education of Taiwan, under contract No. L7091101-19.

Institutional Review Board Statement: Not applicable.

Informed Consent Statement: Not applicable.

Data Availability Statement: Not applicable.

Conflicts of Interest: The authors declare no conflict of interest.

References

1. Crawford, B.; Krayenhoff, E.S.; Cordy, P. The urban energy balance of a lightweight low-rise neighborhood in Andacollo, Chile. *Theor. Appl. Clim.* **2016**, *131*, 55–68. [CrossRef] [PubMed]
2. Rapsomanikis, S.; Trepekli, A.; Loupa, G.; Polyzou, C. Vertical Energy and Momentum Fluxes in the Centre of Athens, Greece During a Heatwave Period (Thermopolis 2009 Campaign). *Boundary-Layer Meteorol.* **2015**, *154*, 497–512. [CrossRef]
3. Akbari, H.; Bretz, S.; Kurn, D.M.; Hanford, J. Peak power and cooling energy savings of high-albedo roofs. *Energy Build.* **1997**, *25*, 117–126. [CrossRef]
4. Zinzi, M.; Fasano, G. Properties and performance of advanced reflective paints to reduce the cooling loads in buildings and mitigate the heat island effect in urban areas. *Int. J. Sustain. Energy* **2009**, *28*, 123–139. [CrossRef]
5. Bozonnet, E.; Doya, M.; Allard, F. Cool roofs impact on building thermal response: A French case study. *Energy Build.* **2011**, *43*, 3006–3012. [CrossRef]
6. Guo, W.; Qiao, X.; Huang, Y.; Fang, M.; Han, X. Study on energy saving effect of heat-reflective insulation coating on envelopes in the hot summer and cold winter zone. *Energy Build.* **2012**, *50*, 196–203. [CrossRef]
7. Garg, V.; Somal, S.; Arumugam, R.; Bhatia, A. Development for cool roof calculator for India. *Energy Build.* **2016**, *114*, 136–142. [CrossRef]
8. Ferrari, C.; Libbra, A.; Cernuschi, F.M.; De Maria, L.; Marchionna, S.; Barozzi, M.; Siligardi, C.; Muscio, A. A composite cool colored tile for sloped roofs with high 'equivalent' solar reflectance. *Energy Build.* **2016**, *114*, 221–226. [CrossRef]
9. Guo, X.; Wang, J.; Wu, Y.; Ao, Y.; Liu, X. Experimental study of the thermal performance of a new type of building reflective coating in hot summer and cold winter zone of China. *Procedia Eng.* **2017**, *205*, 603–608. [CrossRef]
10. Qiu, T.; Wang, G.; Xu, Q.; Ni, G. Study on the thermal performance and design method of solar reflective–thermal insulation hybrid system for wall and roof in Shanghai. *Sol. Energy* **2018**, *171*, 851–862. [CrossRef]
11. Cheng, V.; Ng, E.Y.Y.; Givoni, B. Effect of envelope colour and thermal mass on indoor temperatures in hot humid climate. *Sol. Energy* **2005**, *78*, 528–534. [CrossRef]
12. Uemoto, K.L.; Sato, N.M.; John, V.M. Estimating thermal performance of cool colored paints. *Energy Build.* **2010**, *42*, 17–22. [CrossRef]
13. Del Carpio, J.A.V.; Marinoski, D.L.; Trichês, G.; Lambert, R. Urban pavements used in Brazil: Characterization of solar reflectance and temperature verification in the field. *Sol. Energy* **2016**, *134*, 72–81. [CrossRef]
14. Roman, K.K.; O'Brien, T.; Alvey, J.; Woo, O. Simulating the effects of cool roof and PCM (phase change materials) based roof to mitigate UHI (urban heat island) in prominent US cities. *Energy* **2016**, *96*, 103–117. [CrossRef]
15. Synnefa, A.; Santamouris, M.; Livada, I. A study of the thermal performance of reflective coatings for the urban environment. *Sol. Energy* **2006**, *80*, 968–981. [CrossRef]
16. Carnielo, E.; Fanchiotti, A.; Zinzi, M. Energy and Comfort Benefits of a Cool Roof Application in a Non-Residential Building Belonging to Roma Tre University. *World Renew. Energy Congr.* **2011**, *57*, 1970–1977.
17. Shen, H.; Tan, H.; Tzempelikos, A. The effect of reflective coatings on building surface temperatures, indoor environment and energy consumption—An experimental study. *Energy Build.* **2011**, *43*, 573–580. [CrossRef]
18. Santamouris, M.; Synnefa, A.; Karlessi, T. Using advanced cool materials in the urban built environment to mitigate heat islands and improve thermal comfort conditions. *Sol. Energy* **2011**, *85*, 3085–3102. [CrossRef]
19. Parker, D.S.; Barkaszi, S.F. Roof solar reflectance and cooling energy use: Field research results from Florida. *Energy Build.* **1997**, *25*, 105–115. [CrossRef]
20. Takebayashi, H.; Masakazu, M. Surface heat budget on green roof and high reflection roof for mitigation of urban heat island. *Build. Environ.* **2007**, *42*, 2971–2979. [CrossRef]
21. Costanzo, V.; Evola, G.; Marletta, L. Energy savings in buildings or UHI mitigation? Comparison between green roofs and cool roofs. *Energy Build.* **2016**, *114*, 247–255. [CrossRef]
22. Algarni, S.; Nutter, D. Influence of dust accumulation on building roof thermal performance and radiant heat gain in hot-dry climates. *Energy Build.* **2015**, *104*, 181–190. [CrossRef]
23. Shariah, A.; Shalabi, B.; Rousan, A.; Tashtoush, B. Effects of absorptance of external surfaces on heating and cooling loads of residential buildings in Jordan. *Energy Convers. Manag.* **1998**, *39*, 273–284. [CrossRef]
24. Ruivo, C.; Ferreira, P.; Vaz, D. On the error of calculation of heat gains through walls by methods using constant decrement factor and time lag values. *Energy Build.* **2013**, *60*, 252–261. [CrossRef]
25. *ASTM G173-03. Standard Tables for Reference Solar Spectral Irradiances: Direct Normal and Hemispherical on 37° Tilted Surface*; American Society for Testing and Materials: West Conshohocken, PA, USA, 2020. [CrossRef]
26. Pomianowski, M.; Heiselberg, P.; Jensen, R.L.; Cheng, R.; Zhang, Y. A new experimental method to determine specific heat capacity of inhomogeneous concrete material with incorporated microencapsulated-PCM. *Cem. Concr. Res.* **2014**, *55*, 22–34. [CrossRef]
27. Shafigh, P.; Asadi, I.; Mahyuddin, N.B. Concrete as a thermal mass material for building applications—A review. *J. Build. Eng.* **2018**, *19*, 14–25. [CrossRef]
28. Sarbu, I.; Sebarchievici, C. A Comprehensive Review of Thermal Energy Storage. *Sustainability* **2018**, *10*, 191. [CrossRef]
29. Smith, E.R.; Daivis, P.J.; Todd, B.D. Measuring heat flux beyond Fourier's law. *J. Chem. Physics* **2019**, *150*, 064103. [CrossRef] [PubMed]

30. Childs, P.R.N.; Greenwood, J.R.; Long, C.A. Heat flux measurement techniques. *Proc. Inst. Me Chanical Eng. Part C J. Mech. Eng. Sci.* **1999**, *213*, 655–677. [CrossRef]
31. Misra, K.; Shrotriya, A.K.; Singhvi, N.; Singh, R.; Chaudhary, D.R. Prediction of heat storage coefficient of two-phase systems with spherical inclusions. *J. Phys. D Appl. Phys.* **1994**, *27*, 1823. [CrossRef]
32. Li, M.; Wu, Z.; Tan, J. Heat storage properties of the cement mortar incorporated with composite phase change material. *Appl. Energy* **2013**, *103*, 393–399. [CrossRef]

sustainability

Article

An Experimental Study on Mechanical Behaviors of Carbon Fiber and Microwave-Assisted Pyrolysis Recycled Carbon Fiber-Reinforced Concrete

Yeou-Fong Li [1,*], Jie-You Li [1], Gobinathan Kadagathur Ramanathan [1], Shu-Mei Chang [2], Ming-Yuan Shen [3], Ying-Kuan Tsai [4] and Chih-Hong Huang [5]

1 Department of Civil Engineering, National Taipei University of Technology, Taipei 10608, Taiwan; klsh21505@gmail.com (J.-Y.L.); gobiram0017@gmail.com (G.K.R.)
2 Department of Molecular Science and Engineering, National Taipei University of Technology, Taipei 10608, Taiwan; f10914@mail.ntut.edu.tw
3 Department of Mechanical Engineering, National Chin-Yi University of Technology, Taichung 41170, Taiwan; myshen@ncut.edu.tw
4 Department of Environmental Information and Engineering, Chung Cheng Institute of Technology, National Defense University, Taoyuan 335, Taiwan; jeremytsai0406@gmail.com
5 Department of Architecture, National Taipei University of Technology, Taipei 10608, Taiwan; huangch@mail.ntut.edu.tw
* Correspondence: yfli@mail.ntut.edu.tw

Citation: Li, Y.-F.; Li, J.-Y.; Ramanathan, G.K.; Chang, S.-M.; Shen, M.-Y.; Tsai, Y.-K.; Huang, C.-H. An Experimental Study on Mechanical Behaviors of Carbon Fiber and Microwave-Assisted Pyrolysis Recycled Carbon Fiber-Reinforced Concrete. *Sustainability* **2021**, *13*, 6829. https://doi.org/10.3390/su13126829

Academic Editor: Constantin Chalioris

Received: 25 May 2021
Accepted: 9 June 2021
Published: 17 June 2021

Abstract: In the last decade, waste carbon fiber-reinforced plastic (CFRP) products have not been properly recycled and reused, and they sometimes cause environmental problems. In this paper, the microwave-assisted pyrolysis (MAP) technology was utilized to remove the resin from the CFRP bicycle frame, which was recycled into carbon fiber. A scanning electron microscope (SEM) and single filament tensile test were used to observe and compare the difference between recycled carbon fiber and normal carbon fiber. The mechanical performances of carbon fiber-reinforced concrete (CFRC) were investigated with static and dynamic tests under three different fiber/cement weight proportions (5‰, 10‰, and 15‰). Three different kinds of carbon fiber were used in this study, normal carbon fiber, carbon fiber without coupling agent, and recycled carbon fiber. The experimental program was tested according to ASTM C39-01, ASTM C293, and ACI 544.2R standards for compression, flexural, and impact test, respectively. From the experimental results, addition of 10‰ of carbon fiber into the concrete exhibited maximum compressive and flexural strength. The impact performance of recycled carbon fiber improved the highest impact number compared with normal carbon fiber under different impact energy.

Keywords: recycled carbon fiber; fiber-reinforced concrete; microwave-assisted pyrolysis; shock wave

1. Introduction

1.1. Background

Reinforced concrete structures are deteriorated by seismic and other loadings, which cause structural damage or failure. Recently, carbon fiber-reinforced plastics (CFRPs) have been widely used in structural repair and seismic retrofit. The merits of carbon fiber are that it does not corrode, degrade, fatigue, and it possesses high specific strength. In addition, CFRP is a composite material with carbon fiber as the stiffener and thermosetting or thermoplastic resins as the matrix. Therefore, it has been widely used in the aerospace industry, automotive industry, sports equipment, and civil engineering. However, CFRP is a kind of material that is difficult to decompose by nature and sometimes can produce toxic gases when burned. The traditional disposal of the waste CFRPs is harmful to the environment; thus, an effective recycling method is urgently needed to turn waste CFRPs into potential recycling products, such as the fiber of fiber-reinforced concrete used in civil engineering infrastructures. The world's population is increasing gradually, and it is affecting the environment by raw material wastage.

In the linear economy, renewable and nonrenewable raw material sources are collected and transformed into products until it is discarded as waste. For example, the CFRP bicycle frame has been used in linear economic material and it is discarded as waste. Recently, some carbon fiber composite material has been recycled in different methods, such as mechanical, thermal, and chemical recycling approaches. In this paper, microwave-assisted pyrolysis (MAP) technology approaches are used to remove the resin from the CFRP bicycle frame and transform it into recycled carbon fiber, and then it is used in fiber-reinforced concrete (FRC) structures; the waste materials have been recycled and turn the linear economy to a circular economy. The recycled carbon fiber from the CFRP wastes reduces environmental pollution by being applied in civil engineering.

1.2. Literature Review

Fiber-reinforced cement matrix materials can significantly improve the mechanical properties of fiber-reinforced cement or fiber-reinforced concrete. So far, there have been many studies on fiber-reinforced cement-based materials. Fiber-reinforced concrete can inhibit the formation of large cracks, control the development direction of concrete cracks, and improve the toughness of concrete, thereby improving the tensile and crack resistance of traditional concrete [1,2].

Compared with glass fiber and polypropylene (PP) fiber, steel fiber has better elastic modulus and tensile strength, and the steel fiber-reinforced concrete enhances the effect of shock absorption, compressive and flexural strength of concrete specimens. According to some research test results, adding silica fume can effectively disperse the steel fiber more evenly [3,4]; and adding 1.5% fiber proportion of steel fiber to concrete can increase the impact number in the free-fall impact test [4,5].

There have been many studies on the application of chopped carbon fiber in cement matrix materials over the last few decades. For example, the mechanical properties and microstructures of cement and concrete have been studied by using different types of carbon fibers, carbon fiber lengths, and carbon fiber proportion [6–13]. Furthermore, the level of dispersion of carbon fiber in the cement will greatly affect the strength of the specimen after solidification. Research has shown that different mixing methods will affect the level of dispersion of carbon fiber [14–17]. The carbon fibers were immersed in the hydrolysis method and the furnace heating method to remove silane on the surface of the carbon fiber. The furnace heating method can effectively remove the silane on the surface of the carbon fiber, and combining the furnace heating method with the pneumatic dispersion method can improve the chopped carbon fiber to uniformly disperse inside cement [18].

The amount of industrial waste and commercial waste increases with an increase in demand. Many scholars have begun to study the application of waste in civil engineering to improve the mechanical properties or durability of concrete. Polyethylene terephthalate (PET) waste replacement weight percentage and volume fraction affect the compressive strength. Due to the incorporation of PET waste, the flexural strength and split tensile strength were also deteriorating to the concrete, but the impact resistance had a tiny enhancement effect by weight percentage replacement with aggregates [19,20]. The addition of 1% volume fraction of metalized plastic waste (MPW) fibers into the concrete enhances the compressive strength and impact resistance [21]. The compressive strength was increased by using scrap tire fibers compared with scrap tire fragments. In addition, the fiber-reinforced concrete with a 0.8% weight proportion of recycled tire steel fiber (RTSF) and 0.2% of polypropylene fiber (PPF) had the highest compressive strength. Using PPF instead of RTSF will not greatly affect the flexural strength [22,23]. The glass fiber-reinforced plastics (GFRP) sheets were chopped into short fiber and incorporated into concrete. The recycled GFRP was not affected by alkaline aggregate reaction and shrinkage of concrete. The test results showed that 3 wt % aggregates replacement with GFRP attained maximum compressive and flexural strength. Adding recycled GFRP into concrete improves the impact strength of concrete, but it will reduce the slump flow [24–26]. The chopped recycled CFRP improves the compressive, flexural, and impact strengths of concrete [27–29]. Adding 3 wt % of CFRP waste into the mortar enhances the stiffness and also the flexural strength [29].

Based on the above research background and literature review, this research will use recycled carbon fiber obtained by microwave-assisted pyrolysis (MAP) technology and apply it to fiber-reinforced concrete. The three different types of carbon fibers (recycled carbon fiber, normal carbon fiber, and carbon fiber without coupling agent) were used with different fiber weight proportions (5‰, 10‰, and 15‰). Additionally, the pneumatic dispersion process was used to disperse the fibers uniformly into cement. The mechanical properties of recycled carbon fiber-reinforced concrete were studied through a compression test, three-point bending test, and impact test.

2. Materials

In this study, the mechanical behavior of concrete was enhanced by adding three types of chopped carbon fibers (normal carbon fiber, carbon fiber without coupling agent, and recycled carbon fiber). This section introduces the materials used in the preparation of CFRC, which include the material characteristics of carbon fiber, carbon fiber recycling technology, and coupling agent removal process.

2.1. Carbon Fiber

The lightweight polyacrylonitrile (PAN-based) carbon fibers, which have a high tensile and specific strength, have been applied to the aerospace industry, wind turbine, sports equipment, and automotive parts. The carbon fiber was obtained from Tairylan Division, Formosa Plastics Group; then, the fiber was chopped at Sheng Peng Applied Materials Co., Ltd. [30]. The material properties are listed in Table 1 [31].

Table 1. Material properties of chopped normal carbon fiber.

Material Property	Value
Tensile strength (MPa)	4900
Tensile modulus (GPa)	250
Elongation (%)	2.0
Density (g/cm^3)	1.81
Fiber diameter (μ)	7

2.2. Removal of Coupling Agent on the Surface of Carbon Fiber

The carbon fibers were immersed in pure water for one day, and then we performed GC-MS testing on the immersion solution to identify the type and boiling range of the coupling agent and other substances. The coupling agent was found to be $C_{26}H_{48}O_3Si$ with a molecular weight of 436.74 g/mol under the boiling point of 507.1 ± 50.0 °C at a pressure of 760 mm-Hg [18]. The removal of the coupling agent process is shown in Figure 1; the carbon fibers were wrapped with aluminum foil and then placed into a Muffle furnace (PF-40, Chuan-Hua Precision, New Taipei City, Taiwan) at a high temperature of 550 °C for 3 h.

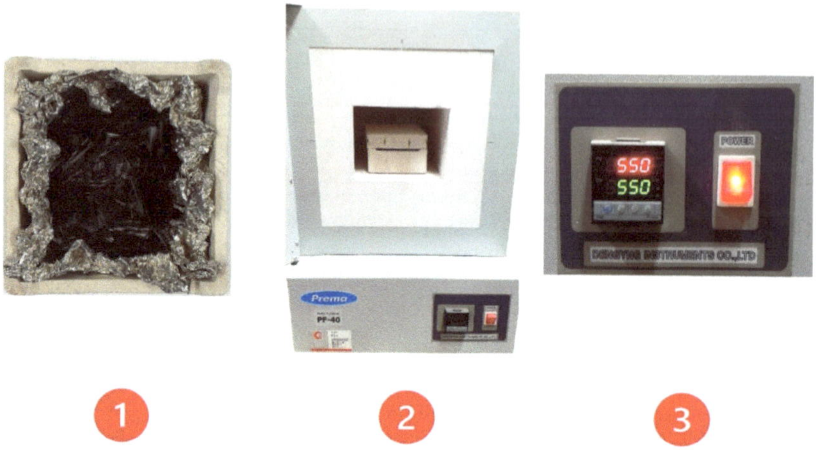

Figure 1. Coupling agent removal processes. 1. Carbon fibers were wrapped with aluminum foil; 2. Placed the carbon fibers with aluminum foil into a Muffle furnace; 3. Set the furnace temperature at 550 °C for 3 h.

Generally, the coupling agent is an inorganic compound, and it reduces the adhesion between the fiber and cement. The differentiation of the coupling agent presence and absence on the surface of carbon fiber is shown in Figure 2. The carbon fiber is easier to distribute in the concrete by removing the coupling agent, compared with normal carbon fiber.

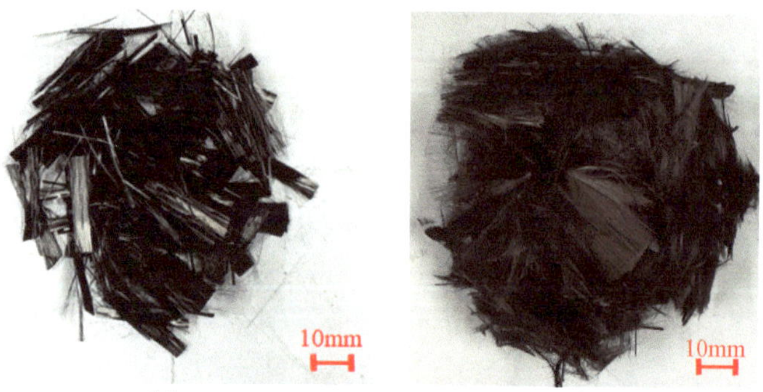

(**a**) (**b**)

Figure 2. The appearance of chopped carbon fiber: (**a**) Normal carbon fiber, (**b**) Carbon fiber without coupling agent.

2.3. Recycled Carbon Fiber

Carbon fiber tow is the thread used to weave carbon fiber fabrics. As a standalone product, it can be used to make wound parts, in pultrusion, or chopped as a local reinforcement. This 12 k tow (or yarn) comprises 12,000 individual carbon filaments, which boast the highest ultimate tensile strength in the industry. The microwave-assisted pyrolysis (MAP) is different from the traditional heating method. It uses microwaves to quickly rub the molecules in the substance, prompting the molecules to rotate quickly to generate heat, and then quickly decompose. The characteristics of microwave heating are: Microwaves only aim at the materials that can absorb them, so energy use is more concentrated. The electric power system is used to generate microwaves, it can be quickly heated, the process is easy to control, and automatic control can be realized. Microwaves penetrate to the absorbable substances, so they can heat the entire object uniformly [32,33].

In the MAP technology, the microwave heat radiations can transfer through the carbon fiber from inner to outer. It can remove the resin from the CFRP bicycle frame scrap to transfer into recycled carbon fiber. Figure 3 shows the microwave machine (PYRO 260, Milestone, Sorisole, Italy) at the Department of Molecular Science and Engineering Laboratory of the National Taipei University of Technology, Taipei, Taiwan. The recycled carbon fiber was obtained from the carbon fiber scrap by using the MAP process in the microwave machine at 950 °C for 1 h, and then a single filament tensile test was carried out.

Figure 3. The microwave machine.

In this study, both carbon fiber and recycled carbon fiber groups were tested in the single filament tensile test and carried out in accordance with ASTM D3379 [34]. Each group of carbon fibers had 30 individual filaments. The load–displacement relationships of the normal and recycled carbon fibers are shown in Figure 4.

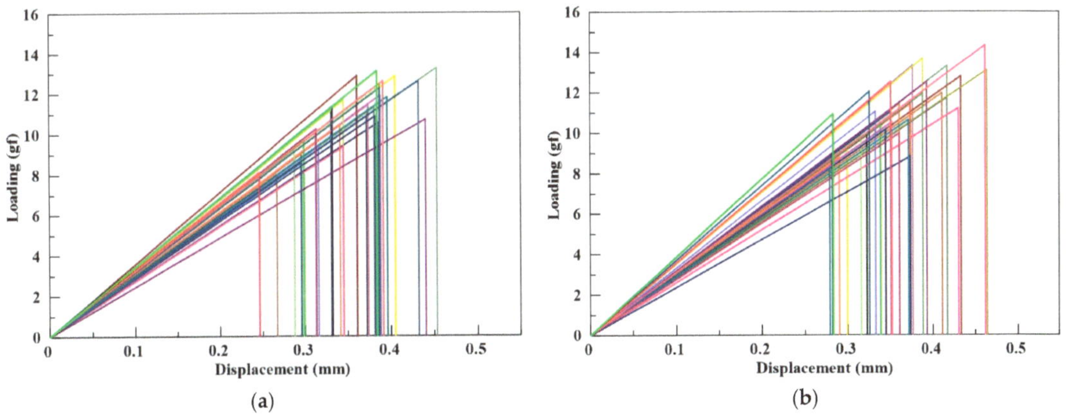

Figure 4. The load–displacement curves of the single filament tensile test: (**a**) Normal carbon fiber, (**b**) Recycled carbon fiber.

The single filament tensile test results show that the tensile strength of recycled carbon fiber treated with MAP is similar to that of the normal carbon fiber. This proves that MAP technology can effectively recycle CFRP waste and retain its mechanical strength.

2.4. SEM Surface Morphology of Carbon Fiber

In order to understand the cleanliness of the resin on the surface of the recycled carbon fiber, in addition to observing the samples before and after the MAP treatment with a scanning electron microscope (SEM), energy dispersive X-ray spectrometry (EDS) was also used to observe the amount of removal. Figure 5a–d shows the photos and SEM images of the waste (recycled) CFRP before and after MAP technology. Using EDS to measure the carbon content of the carbon fiber, the carbon content of the recycled carbon fiber after MAP technology was 99.8%. The SEM image is shown in Figure 5d.

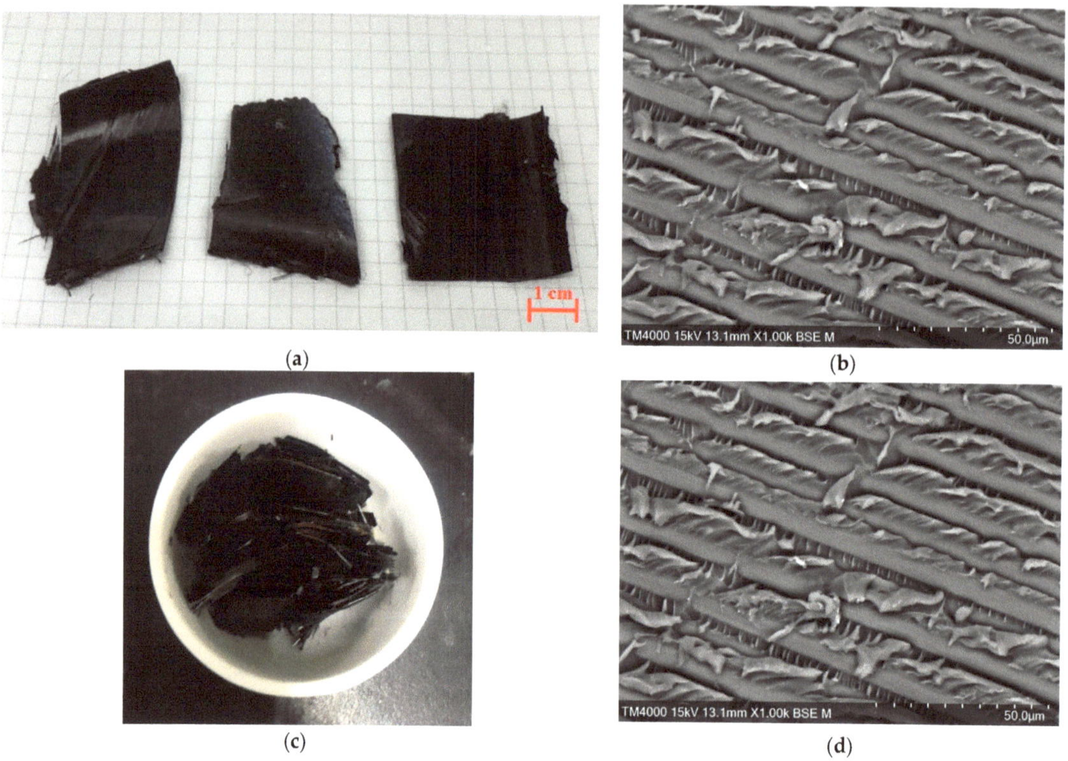

Figure 5. Photos and SEM images of fragment samples of recycled carbon fiber composite materials. (**a**) Fragment samples of recycled CFRP before MAP technology; (**b**) SEM image before MAP technology; (**c**) fragment samples of recycled CFRP after MAP technology; (**d**) SEM image after MAP technology.

The surface morphology of the chopped carbon fiber with and without coupling agent was analyzed by scanning electron microscope (model: JSM-7610F, JEOL, Tokyo, Japan) in the Department of Molecular Science and Engineering Laboratory of the National Taipei University of Technology, Taipei, Taiwan. The surfaces of normal carbon fiber (with coupling agent) and carbon fiber without coupling agent were observed, and their SEM images are shown in Figure 6. Using EDS to measure the carbon content of the carbon fiber, the carbon contents of the normal carbon fiber with and without furnace heating process were 100% and 99.2%, respectively. The coupling agent on the surface of the carbon fiber might have interfered with the bonding strength between the carbon fiber and cement by the furnace heating method.

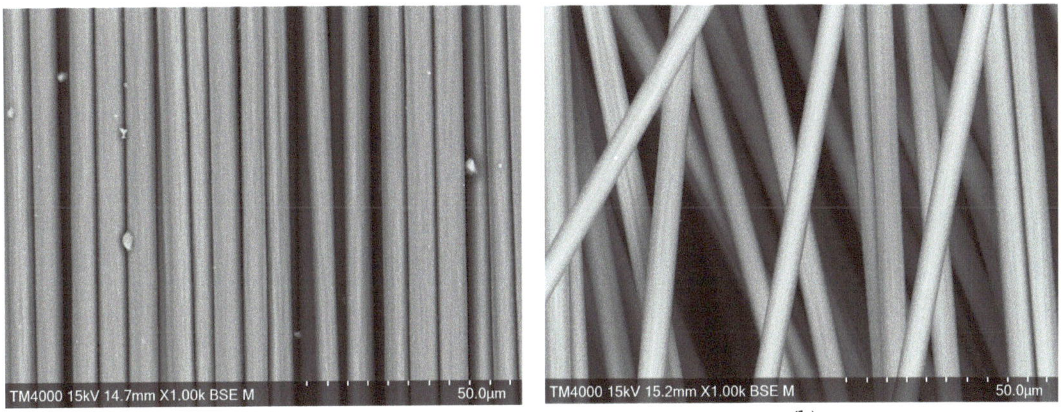

Figure 6. SEM observation results of the surface of normal carbon fiber and carbon fiber without coupling agent. (**a**) SEM image of the chopped carbon fiber without furnace heating; (**b**) SEM image of the chopped carbon fiber with furnace heating.

2.5. Carbon Fiber-Reinforced Concrete

Carbon fiber is a non-corrodible material. The CFRC can reduce the cracks during the service life of concrete structures, it can preserve the steel rebar from corrosions, and mitigate impact loading. In this study, Portland cement was obtained from the Taiwan Cement Corporation [35]. The concrete (benchmark) and CFRC specimens were tested under compressive, three-point bending, and impact tests. The concrete water–cement ratio was 0.6; the cement, sand, fine aggregate, and coarse aggregate ratio was 1:1.05:1.5:0.75. The fineness modulus of the fine aggregate (3/8″) and coarse aggregate (6/8″) were 3.03 and 7.33, respectively. The fineness modulus (F.M.) of both aggregates for the concrete specimen was 6.01, as shown in Table 2. The concrete and CFRC specimens were cured at 28 days.

Table 2. Fineness modulus of aggregates.

Sieve No.	Weight Retained (g)	Percent Retained (%)	Cumulative Percent Retained (%)
3/2″	0	0	0
3/4″	672.3	23	23
3/8″	1352.4	46.2	69.2
No. 4	10.2	0.3	69.5
No. 8	165.6	5.7	75.2
No. 16	236.7	8.1	83.3
No. 30	178.2	6.1	89.4
No. 50	146.7	5	94.4
No. 100	83.7	2.9	97.3
Pan	79.2	2.7	100
Total	2925	-	Cumulative = 6.01

3. Experimental Methods and Setups

In this study, the CFRC specimens were prepared using three kinds of carbon fiber (recycled carbon fiber, normal carbon fiber, and carbon fiber without coupling agent). The compressive test, flexural test, and impact test followed the ASTM and ACI standards.

3.1. Experiment Planning

The three types of carbon fiber (normal carbon fiber, recycled carbon fiber, and carbon fiber without coupling agent) were planned to be used to prepare CFRC specimens with three different fiber weight proportions (5‰, 10‰, and 15‰), and the specimen names and descriptions, and the planning of the specimens are shown in Tables 3 and 4, respectively. For example, specimen C-R05 represents the compression test of recycled carbon fiber with the addition of 5‰ weight proportion.

Table 3. The naming and descriptions of CFRC specimens.

Naming	Description
C	compressive test
F	flexural test
I	impact test
B	benchmark (without carbon fiber)
N	carbon fiber with coupling agent
W	carbon fiber without coupling agent
R	recycled carbon fiber
Weight proportion (‰)	5, 10, 15

Table 4. Planning of CFRC specimens.

Experiment	Fiber Weight Proportion (‰)	Carbon Fiber			Benchmark	Total
		Normal	Recycled	Without Coupling Agent		
Compressive test	5	3	3	3	3	30
	10	3	3	3		
	15	3	3	3		
Flexural test	5	3	3	3	3	30
	10	3	3	3		
	15	3	3	3		
Impact test	10	25	25	25	25	100

3.2. Slump Test

The workability of the fiber-reinforced concrete has a significant impact on the construction quality, and good workability will prevent honeycombs in the concrete that reduce its strength. In this study, the slumps of CFRC of different types of carbon fiber and different weight proportions were tested according to ASTM C143/C143M−20; the slump fluidity range was about 15–230 mm, respectively [36].

3.3. Compressive Test

To explore the compressive strength of the three types of CFRC with different fiber additions, the dispersed chopped carbon fiber was mixed into concrete at a cement weight ratio of 5‰, 10‰, and 15‰. The diameter of the CFRC cylindrical specimen was 10 cm and the height was 20 cm. According to ASTM C39/C 39M-01 [37], the universal testing machine (HT-9501 Series. Hong-Ta, Taipei, Taiwan) was used for testing.

3.4. Three-Point Bending Test

According to ASTM C293-02 [38], the bending test of 28 cm × 7 cm × 7 cm CFRC specimens was carried out by a universal testing machine (HT-9501 Series. Hong-Ta, Taipei, Taiwan) with a load cell (WF 17120, Wykeham Farrance, Milan, Italy). The bending specimen was tested in the material laboratory of the Department of Civil and Disaster Prevention Engineering, National Taipei University of Technology. The three-point bending test is shown in Figure 7.

Figure 7. CFRC specimen flexural test setup.

3.5. Impact Test

According to ACI 544-2R [39], the impact test of ϕ150 mm × 64 mm CFRC specimens was carried out by impact equipment (SP-006, Sheng Peng, Yunlin, Taiwan). The impact energy was 25 J as the interval, and the number of repeated impacts was measured from 50 to 150 J. Figure 8a shows the impact specimen and impact device, and Figure 8b shows the impact test equipment.

(**a**) (**b**)

Figure 8. Impact test: (**a**) CFRC with impact device, (**b**) impact test equipment.

4. Experimental Results and Discussions

Three types of CFRC specimens were prepared from different fibers (recycled carbon fiber, normal carbon fiber, and carbon fiber without coupling agent). The test results of compressive, bending, and impact performance were obtained with the different kinds of carbon fiber in CFRC specimens. The recycled carbon fibers were fabricated by a scrap of CFRP bicycle frame with a MAP technology. The recycled carbon fiber was adopted from Thermolysis Co., Ltd. (Kao-Hsiung, Taiwan) [40]. The recycled carbon fiber was chopped to a length between 20 and 30 mm. The microwave conditions were established according to the process in Section 2.3.

4.1. Slump Test Result

The workability depends on the w/c ratio and weight proportions of fibers. The slump values of CFRC with different types of carbon fiber and different weight proportions are shown in Table 5. The test results showed that the slump value was not affected by the types of carbon fibers but affected by different weight proportions. The recycled carbon fiber with a length of 20–30 mm had a slight variation in the slump according to carbon fiber coupling agent presence and absence. The CFRC with 5‰ fiber weight proportion of the carbon fiber had the best workability, and its slump was about 160 mm. The CFRC with 15‰ fiber weight proportion of the carbon fiber became stickier, making the CFRC wet mixing hard to mix and sometimes diminished its mechanical strength. The CFRC mixture with more than 15‰ weight proportion had less workability under 0.60 w/c ratio.

Table 5. Slump test of different fiber weight proportions.

Fiber Weight Proportion (‰)	Slump of CFRC (mm)		
	Recycled	Normal	Carbon Fiber without Coupling Agent
0	230	230	230
5	160	165	165
10	85	80	80
15	40	40	40

4.2. Compressive Test Result

In this study, the three different kinds of carbon fiber-reinforced concrete were subjected to compression test by uniaxial loading. Figure 9 shows the average compressive strength of CFRC and benchmark under different proportions. Compared with other fiber weight proportions, adding 10‰ fiber weight proportion of CFRC can increase the maximum compressive strength.

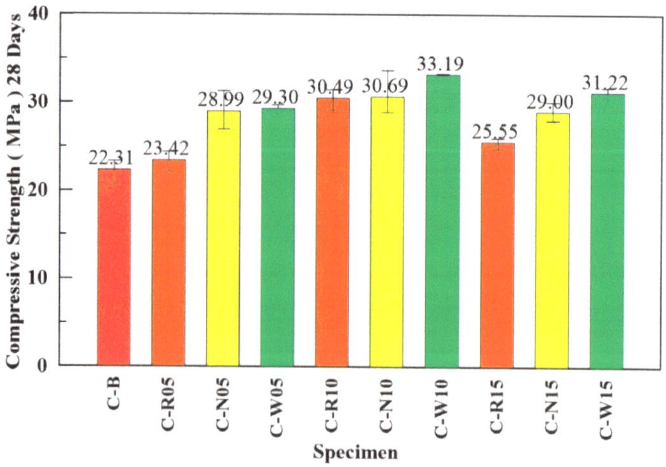

Figure 9. Average compressive strength of CFRC and benchmark specimens. (Note: C—Compression test; B—Benchmark; R—Recycled carbon fiber; N—Normal carbon fiber; W—Carbon fiber without coupling agent).

Table 6 shows the compressive strength of different CFRC and benchmark under different fiber weight proportions. Under three different carbon fiber weight proportions, the recycled carbon fiber-reinforced concrete and normal carbon fiber-reinforced concrete did not exhibit a greater enhancement effect on compressive strength than the carbon fiber-reinforced concrete by removing the coupling agent. For instance, the C-W10 specimen exhibited the highest compressive strength (33.19 MPa) compared with C-B specimen, C-N10 (30.69 MPa), and C-R10 (30.49 MPa), as shown in Table 6.

Table 6. Compressive strengths of CFRC and benchmark under different proportions.

Specimen	Number	Compressive Strength (MPa)	Average (MPa)	Increase (%)
C-B	1	20.60	22.31	-
	2	23.32		
	3	22.99		
C-R05	1	22.19	23.42	4.9
	2	23.69		
	3	24.39		
C-N05	1	26.94	28.99	29.9
	2	31.32		
	3	28.72		
C-W05	1	29.49	29.30	31.3
	2	28.68		
	3	29.72		
C-R10	1	30.87	30.49	36.7
	2	31.47		
	3	29.14		
C-N10	1	33.65	30.69	37.6
	2	29.50		
	3	28.91		
C-W10	1	33.27	33.19	48.9
	2	33.15		
	3	33.15		
C-R15	1	25.84	25.55	14.5
	2	26.01		
	3	24.81		
C-N15	1	30.08	29.00	30
	2	28.93		
	3	28.00		
C-W15	1	31.73	31.22	39.9
	2	31.42		
	3	30.52		

Note: C—Compression test; B—Benchmark; R—Recycled carbon fiber; N—Normal carbon fiber; W—Carbon fiber without coupling agent.

When the fiber weight proportion was 10‰, the compressive strength of specimen C-W10 was the highest, followed by specimen C-N10 and specimen C-R10. Using EDS to measure the carbon content of the carbon fiber, the carbon contents of the normal carbon fiber with the furnace heating process, the recycled carbon fiber with the MAP process, and the normal carbon fiber were 100, 99.8, and 99.2%, respectively. The test results showed that the higher the carbon content of the carbon fibers, the higher the compressive strength of the CFRC specimen. The surface of the carbon fiber without coupling agent or resin residual, the adhesion force between carbon fiber and cement increased. Therefore, the CFRC specimen without coupling agent (C-W10) has the highest compressive strength.

4.3. Three-Point Bending Test Result

In this Subsection, the flexural strength of three kinds of CFRC with different fiber weight proportions are compared with the benchmark specimens. As shown in Figure 10, the CFRC with 10‰ of fiber weight proportion increased its flexural strength more than the other fiber weight proportions, such as 5‰ and 15‰. From the slump test results, the 15‰ fiber weight proportion of CFRC was close to the lower range compared with the ASTM C143/C143M-20 standard requirement (mention in Section 3.2), respectively. Therefore, the carbon fibers were not easy to distribute uniformly in the FRC, and the flexural strength was reduced. Additionally, the CFRC with 5‰ fiber weight proportion of carbon fiber was not enough to increase the flexural strength because the amount of carbon fiber was too low.

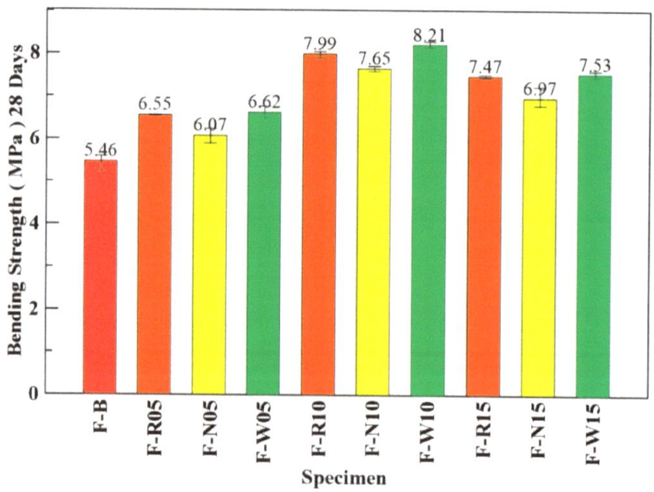

Figure 10. Average flexural strengths of CFRC and benchmark specimens. (Note: F—Flexural; B—Benchmark; R—Recycled carbon fiber; N—Normal carbon fiber; W—Carbon fiber without coupling agent).

Table 7 shows the flexural strength of CFRC and benchmark under different fiber weight proportions. The F-W10 specimen increased its flexural strength up to 50.4% compared with F-B specimen. Similarly, the flexural strength of F-R10 and F-N10 had higher strength than the benchmark specimen at 46.3 and 39.6%, respectively. In Section 2.4, the SEM images with the corresponding EDS showed the carbon content of carbon fiber without coupling agent, recycled carbon fiber, and normal carbon fiber were 100, 99.8, and 99.2%, respectively. As seen from the flexural test results and the EDS measurement, the flexural strength increased with the carbon contents of the carbon fiber; the fewer residuals increased the flexural strength of the CFRC specimens.

Table 7. Flexural strengths of CFRC and benchmark under different proportions.

Specimen	Number	Compressive Strength (MPa)	Average (MPa)	Increase (%)
F-B	1	5.55	5.46	-
	2	5.60		
	3	5.24		
F-N05	1	5.90	6.07	11.2
	2	6.24		
	3	6.08		
F-R05	1	6.55	6.55	20.0
	2	6.54		
	3	6.56		
F-W05	1	6.77	6.62	21.2
	2	6.50		
	3	6.58		
F-N10	1	7.65	7.62	39.6
	2	7.59		
	3	7.72		
F-R10	1	8.00	7.99	46.3
	2	7.90		
	3	8.06		
F-W10	1	8.29	8.21	50.4
	2	8.18		
	3	8.16		
F-N15	1	7.24	6.97	27.7
	2	6.86		
	3	6.80		
F-R15	1	7.46	7.47	36.6
	2	7.51		
	3	7.45		
F-W15	1	7.52	7.53	37.9
	2	7.61		
	3	7.46		

Note: F—Flexural; B—Benchmark; R—Recycled carbon fiber; N—Normal carbon fiber; W—Carbon fiber without coupling agent).

4.4. Impact Test Result

The impact numbers of benchmark and CFRC specimens under different impact energies are shown in Table 8. From the test results, the I-N10, I-R10, I-W10 specimens resisted repeated impact at low energy of 50 J and the average impact numbers were about 285, 356, and 410, respectively. Compared with the I-B specimen, the I-N10, I-R10, and I-W10 specimens and the impact number increase percentages were about 1828%, 2305% and, 2669%, respectively. From the impact test result, the CFRC specimens with recycled carbon fiber and carbon fiber without coupling agent exhibited enhancement effects compared to the CFRC specimen with normal carbon fiber.

Table 8. Impact test results of CFRC specimens under different impact energies.

Specimen	Impact Energy (J)	Specimen Number					Average Impact Number	Increase Percentage (%)
		1	2	3	4	5		
I-B	150	1	1	1	1	1	1.0	-
	125	2	2	2	2	3	2.2	-
	100	3	3	4	4	4	3.6	-
	75	5	6	6	7	7	6.2	-
	50	13	14	15	16	16	14.8	-
I-N10	150	1	1	1	2	2	1.4	40.0
	125	3	3	4	4	4	3.6	63.6
	100	10	11	13	14	15	12.6	250.0
	75	53	60	62	63	66	60.8	880.6
	50	267	275	288	293	304	285.4	1828.4
I-R10	150	1	1	2	2	2	1.6	60.0
	125	3	4	4	4	4	3.8	72.7
	100	16	17	18	19	19	17.8	394.4
	75	74	76	87	88	91	83.2	1241.9
	50	338	349	353	366	374	356	2305.4
I-W10	150	1	2	2	2	2	1.8	80
	125	3	4	4	4	5	4.0	81.8
	100	16	16	18	19	21	18	400.0
	75	78	81	86	92	98	87	1303.2
	50	385	392	409	429	434	409.8	2668.9

Note: I—Impact; B—Benchmark; N—normal carbon fiber; R—recycled carbon fiber; W—carbon fiber without coupling agent.

Figure 11 shows the impact energy/number curve of CFRC specimens. It can be clearly seen that the repeated impact number of CFRC specimens under low impact energy is greater than that under high impact energy.

As seen in Figure 11, the I-W10 specimen had the highest average number of repeated impacts when the impact energy was 50 J, followed by I-R10 and I-N10. The EDS measurement results and impact test results show that the higher carbon contents of the carbon fiber had higher repeated impact capability of CFRC.

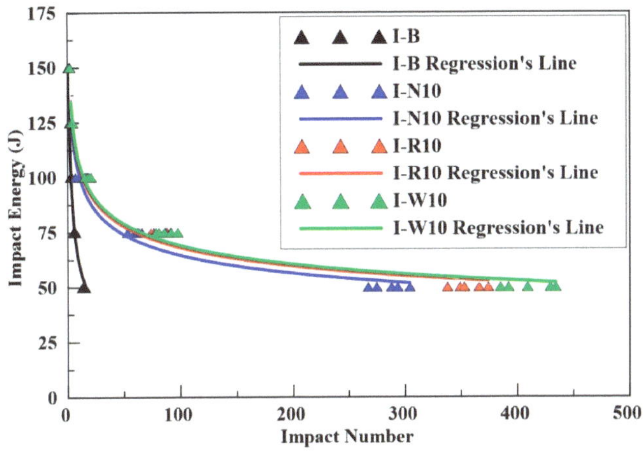

Figure 11. The curves of impact energy/number of CFRC specimens. (Note: I—Impact; B—Benchmark; N—Normal carbon fiber; R—Recycled carbon fiber; W—Carbon fiber without coupling agent).

5. Conclusions

Regarding the CFRC with different proportions of normal carbon fiber, recycled carbon fiber, and carbon fiber without coupling agent under static and impact tests several conclusions can be stated as follows.

1. According the SEM-EDS, the carbon content on the surface of recycled carbon fiber was about 99.8% with the MAP technology process, which effectively removed the resin to revert the carbon fiber of the CFRP product. In addition, the carbon content for normal carbon fiber and carbon fiber without coupling agent were about 99.2 and 100%, respectively.
2. The test results showed that the slump value was not affected by the carbon fibers but affected by different weight proportions.
3. The 10‰ proportion carbon fiber without coupling agent exhibited maximum compressive strength compared with the benchmark, recycled, and normal carbon fiber. The maximum compressive strengths of the C-W10, C-R10, and C-N10 specimens were compared with the C-B specimen, and the increased percentages were about 48.8%, 37.6%, and 36.7%.
4. The CFRC with 10‰ fiber weight proportion increased its flexural strength compared with other proportions because the 5% was not enough to increase the strength compared with 10%, and the 15% proportions were too close to the lowest slump value (mm) in the standard. The high fiber weight proportion made the CFRC difficult to mix and reduced its strength.
5. From the impact test, the CFRC with the absence of coupling agent (I-W10) had high impact resistance, and then recycled (I-R10) had a high impact resistance number compared to carbon fiber with the presence of coupling agent (I-N10) under different energies (J).
6. According to the test results, the flexural strengths and impact resistances of the CFRC specimens were closely related to the carbon content of carbon fiber measured by EDS. We saw from the experiment that the higher carbon content had higher flexural strength and higher impact resistance.
7. The mechanical performance of recycled carbon fiber was superior to the normal carbon fiber, and it was almost similar to the carbon fiber-reinforced concrete by removal of silane. Using the MAP approaches, the waste scrap of CFRP can be

recycled to carbon fibers and applied to reinforced structures such as bridge expansion joints, tunnels, dams, airport and highway pavements, etc.

Author Contributions: Conceptualization, Y.-F.L.; data curation, J.-Y.L. and G.K.R.; formal analysis, J.-Y.L. and G.K.R.; investigation, J.-Y.L., and G.K.R.; methodology, Y.-F.L., S.-M.C. and M.-Y.S.; project administration, Y.-F.L. and C.-H.H.; supervision, Y.-F.L., S.-M.C. and Y.-K.T.; writing—original draft, J.-Y.L., and G.K.R.; writing—review and editing, Y.-F.L., S.-M.C., M.-Y.S., Y.-K.T. and C.-H.H. All authors have read and agreed to the published version of the manuscript.

Funding: This research was funded by the Ministry of Science and Technology of Taiwan government, under contract No. MOST—108-2218-E-027—005, and the "Research Center of Energy Conservation for New Generation of Residential, Commercial, and Industrial Sectors" from the Ministry of Education in Taiwan under contract No. L7091101-19.

Institutional Review Board Statement: Not applicable.

Informed Consent Statement: Not applicable.

Data Availability Statement: Not applicable.

Conflicts of Interest: The authors declare no conflict of interest.

References

1. Erdem, S.; Hanbay, S.; Blankson, M.A. Self-sensing damage assessment and image-based surface crack quantification of carbon nanofiber reinforced concrete. Constr. *Build. Mater.* **2017**, *134*, 520–529. [CrossRef]
2. Li, Y.; Li, Y. Experimental study on performance of rubber particle and steel fiber composite toughening concrete. *Constr. Build. Mater.* **2017**, *146*, 267–275. [CrossRef]
3. Giner, V.T.; Baeza, F.J.; Ivorra, S.; Zornoza, E.; Galao, O. Effect of steel and carbon fiber additions on the dynamic properties of concrete containing silica fume. *Mater. Des.* **2012**, *34*, 332–339. [CrossRef]
4. Nili, M.; Afroughsabet, V. Combined effect of silica fume and steel fibers on the impact resistance and mechanical properties of concrete. *Int. J. Impact Eng.* **2010**, *37*, 879–886. [CrossRef]
5. Naraganti, S.R.; Pannem, R.M.R.; Putta, J. Impact resistance of hybrid fibre reinforced concrete containing sisal fibres. *Ain Shams Eng. J.* **2019**, *10*, 297–305. [CrossRef]
6. Chen, P.W.; Chung, D.D.L. Concrete reinforced with up to 0.2 vol% of short carbon fibres. *Composites* **1993**, *24*, 33–52. [CrossRef]
7. Li, V.C.; Obla, K.H. Effect of fiber length variation on tensile properties of carbon-fiber cement composites. *Compos. Eng.* **1994**, *4*, 947–964. [CrossRef]
8. Park, S.-J.; Seo, M.-K.; Shim, H.-B.; Rhee, K.Y. Effect of different cross-section types on mechanical properties of carbon fibers-reinforced cement composites. *Mater. Sci. Eng. A* **2014**, *366*, 348–355. [CrossRef]
9. Tabatabaei, Z.S.; Volz, J.S.; Keener, D.I.; Gliha, B.P. Comparative impact behavior of four long carbon fiber reinforced concretes. *Mater. Des.* **2014**, *55*, 212–223. [CrossRef]
10. Rangelov, M.; Nassiri, S.; Haselbach, L.; Englund, K. Using carbon fiber composites for reinforcing pervious concrete. *Constr. Build. Mater.* **2016**, *126*, 875–885. [CrossRef]
11. Meng, W.; Khayat, K.H. Mechanical properties of ultra-high-performance concrete enhanced with graphite nanoplatelets and carbon nanofibers. *Compos. B Eng.* **2016**, *107*, 113–122. [CrossRef]
12. Han, B.; Zhang, L.; Zhang, C.; Wang, Y.; Yu, X.; Ou, J. Reinforcement effect and mechanism of carbon fibers to mechanical and electrically conductive properties of cement-based materials. *Constr. Build. Mater.* **2016**, *125*, 479–486. [CrossRef]
13. Li, Y.-F.; Lee, K.-F.; Kadagathur Ramanathan, G.; Cheng, T.-W.; Huang, C.-H.; Tsai, Y.-K. Static and Dynamic Performances of Chopped Carbon-Fiber-Reinforced Mortar and Concrete Incorporated with Disparate Lengths. *Materials* **2021**, *14*, 972. [CrossRef] [PubMed]
14. Xu, Y.; Chung, D. Carbon fiber reinforced cement improved by using silane-treated carbon fibers. *Cem. Concr. Res.* **1999**, *29*, 773–776. [CrossRef]
15. Yang, Y. Methods Study on Dispersion of Fibers in CFRC. *Cem. Concr. Res.* **2001**, *32*, 747–750. [CrossRef]
16. Wang, Z.; Gao, J.; Ai, T.; Jiang, W.; Zhao, P. Quantitative evaluation of carbon fiber dispersion in cement based composites. *Constr. Build. Mater.* **2014**, *68*, 26–30. [CrossRef]
17. Gao, J.; Wang, Z.; Zhang, T.; Zhou, L. Dispersion of carbon fibers in cement-based composites with different mixing methods. *Constr. Build. Mater.* **2017**, *134*, 220–227. [CrossRef]
18. Li, Y.-F.; Yang, T.-H.; Kuo, C.-Y.; Tsai, Y.-K. A Study on Improving the Mechanical Performance of Carbon-Fiber-Reinforced. *Cement. Mater.* **2019**, *12*, 2715. [CrossRef] [PubMed]
19. Saxena, R.; Siddique, S.; Gupta, T.; Sharma, R.K.; Chaudhary, S. Impact resistance and energy absorption capacity of concrete containing plastic waste. *Constr. Build. Mater.* **2018**, *176*, 415–421. [CrossRef]

20. Saikia, N.; de Brito, J. Mechanical properties and abrasion behaviour of concrete containing shredded PET bottle waste as a partial substitution of natural aggregate. *Constr. Build. Mater.* **2014**, *52*, 236–244. [CrossRef]
21. Ankur, C.; Narendra, K. Impact strength, permeability and chemical resistance of concrete reinforced with metalized plastic waste fibers. *Constr. Build. Mater.* **2018**, *161*, 254–266.
22. Li, G.; Stubblefield, M.; Garrick, G.; Eggers, J.; Abadie, C.; Huang, B. Development of waste tire modified concrete. *Cem. Concr. Res.* **2004**, *34*, 2283–2289. [CrossRef]
23. Zhong, H.; Zhang, M. Experimental study on engineering properties of concrete reinforced with hybrid recycled tyre steel and polypropylene fibres. *J. Clean. Prod.* **2020**, *259*, 120914. [CrossRef]
24. García, D.; Vegas, I.; Cacho, I. Mechanical recycling of GFRP waste as short-fiber reinforcements in microconcrete. *Constr. Build. Mater.* **2014**, *64*, 293–300. [CrossRef]
25. Rodin III, H.; Nassiri, S.; Englund, K.; Fakron, O.; Li, H. Recycled glass fiber reinforced polymer composites incorporated in mortar for improved mechanical performance. *Constr. Build. Mater.* **2018**, *187*, 738–751. [CrossRef]
26. Mastali, M.; Dalvand, A.; Sattarifard, A. The impact resistance and mechanical properties of reinforced self-compacting concrete with recycled glass fibre reinforced polymers. *J. Clean. Prod.* **2016**, *124*, 312–324. [CrossRef]
27. Ogi, K.; Shinoda, T.; Makoto, M. Strength in concrete reinforced with recycled CFRP pieces. *Compos. A Appl. Sci.* **2005**, *36*, 893–902. [CrossRef]
28. Mastali, M.; Dalvand, A. The impact resistance and mechanical properties of self-compacting concrete reinforced with recycled CFRP pieces. *Compos. B Eng.* **2016**, *92*, 360–376. [CrossRef]
29. Nguyen, H.; Fujii, T.; Okubo, K.; Carvelli, V. Cement mortar reinforced with recycled carbon fiber and CFRP waste. In Proceedings of the ECCM17—17th European Conference on Composite Materials, München, Germany, 26–30 June 2016.
30. Sheng Peng Applied Materials Co., Ltd. Available online: http://www.spco.com.tw/index.aspx?lang=US (accessed on 18 March 2021).
31. Tairylan Division, FPG. Available online: http://www.jeccomposites.com/directory/formosa-plastics-corporation (accessed on 8 January 2021).
32. Yin, C. Microwave-assisted pyrolysis of biomass for liquid biofuels production. *Bioresour. Technol.* **2012**, *120*, 273–284. [CrossRef]
33. Motasemi, F.; Afzal, M.T. A review on the microwave-assisted pyrolysis technique. *Renew. Sustain. Energy Rev.* **2013**, *28*, 317–330.
34. ASTM D3379. *Standard Test Method for Tensile Strength and Young's Modulus for High-Modulus Single-Filament Materials*; ASTM Annual Book of Standards; ASTM: West Conshohocken, PA, USA, 1989; pp. 128–131.
35. Taiwan Cement Corporation. Available online: https://www.taiwancement.com/en/aboutProduct.html (accessed on 22 February 2021).
36. ASTMC143/C143M-20. *Standard Test Method for Slump of Hydraulic–Cement Concrete*; ASTM: West Conshohocken, PA, USA, 2015.
37. ASTM C39/C39M-01. *Standard Test Method for Compressive Strength of Cylindrical Concrete Specimens*; ASTM: West Conshohocken, PA, USA, 2012.
38. ASTM C293-02. *Standard Test Method for Flexural Strength of Concrete (Using Simple Beam with Center-Point Loading)*; ASTM: West Conshohocken, PA, USA, 2010.
39. ACI-544-2R89. *Measurement of Properties of Fiber Reinforced Concrete*; America Concrete Institute: Farmington Hills, MI, USA, 1999; pp. 6–7.
40. Thermolysis Co., Ltd. Available online: https://www.thermolysis-asia.com/ (accessed on 18 March 2021).

sustainability MDPI

Article

A Study on Radiation Cooling Effect on Asphalt Concrete Pavement Using Basic Oxygen Furnace Slag to Replace Partial Aggregates

Yeou-Fong Li [1,*], Po-An Yang [1], Chia-Ho Wu [2], Ta-Wui Cheng [2] and Chih-Hong Huang [3,*]

1 Department of Civil Engineering, National Taipei University of Technology, 1, Sec. 3, Chung-Hsiao E. Rd., Taipei 10608, Taiwan; 001068smhs@gmail.com
2 Institute of Mineral Resources Engineering, National Taipei University of Technology, Taipei 10608, Taiwan; cep03210@gmail.com (C.-H.W.); twcheng@ntut.edu.tw (T.-W.C.)
3 Department of Architecture, National Taipei University of Technology, Taipei 10608, Taiwan
* Correspondence: yfli@mail.ntut.edu.tw (Y.-F.L.); huangch@mail.ntut.edu.tw (C.-H.H.)

Abstract: In this study, aggregates in asphalt concrete were partially replaced by basic oxygen furnace slag (BOFS) in proportions of 45 wt.%, 55 wt.%, and 75 wt.%. The thermal performances of the specimens are discussed based on the thermal conductivity, emissivity, and the indoor and outdoor temperature measurements. Consequently, 75 wt.% of the specimen's aggregates were replaced by BOFS, which had a high emissivity of 0.86 across the sky window. In the indoor and outdoor tests, the temperature change was recorded to estimate the thermal performance of specimens. According to the quantitative calculation, when the substitution of BOFS was higher than 55 wt.%, the specimens had a better radiation cooling ability. Among these specimens, the specimen with the BOFS substitution of 75 wt.% absorbed the most heat inside the body, contributing to less heat remaining in the environment. Furthermore, because Newton's cooling energy accounted for about 90% of the stored energy within 7 h, the heat dissipation after the seventh hour was primarily radiation cooling, corresponding to the emission across the urban boundary layer. As for the mechanical properties, the stability value, indirect tensile strength, and British pendulum number (BPN) were in line with the specifications under the proper BOFS substitution. In conclusion, BOFS has great applicability in pavements due to its thermal performance and mechanical properties. It not only achieves the goal of urban heat island mitigation by radiation cooling, but also reflects the concept of resource sustainability.

Keywords: asphalt concrete; basic oxygen furnace slag; radiation cooling; emissivity; thermal conductivity

Citation: Li, Y.-F.; Yang, P.-A.; Wu, C.-H.; Cheng, T.-W.; Huang, C.-H. A Study on Radiation Cooling Effect on Asphalt Concrete Pavement Using Basic Oxygen Furnace Slag to Replace Partial Aggregates. *Sustainability* **2021**, *13*, 3708. https://doi.org/10.3390/su13073708

Academic Editor: Jorge de Brito

Received: 22 February 2021
Accepted: 23 March 2021
Published: 26 March 2021

1. Introduction

In this century with booming industries, the enormous exploitation and consumption of energy results in the ever-increasing concentration of carbon dioxide worldwide, causing the greenhouse effect and extreme climate changes. According to the statistics from the National Oceanic and Atmospheric Administration (NOAA), the average temperature has increased by up to 0.18 °C per decade since 1981. This has caused the snow in the polar regions to melt and subsequently raised the sea level [1]. Furthermore, the concentration of carbon dioxide started at about 265 ppm in 1850, reached 385 ppm in 2009, and even ascended to 407.4 ppm in 2018. The vicious circle of greenhouse gas emission has a tremendous impact not only on metropolises but also environments. Under the circumstances, building envelopes and pavements with high heat capacities bring on and even worsen the urban heat island effect.

Nowadays, to mitigate the urban heat island effect, many studies have proposed to utilize reflective coating of asphalt concrete and have recorded its ambient temperatures and internal temperatures at different depths, exploring the principle of cooling with its spectral characteristics [2,3]. One study changed the color of asphalt concrete to discuss its reflection characteristics [4]; some even improved the reflectivity across specific wavelengths, mainly aimed at visible and near infrared light [5–7]; others made multi-layer structure coatings and compared their cooling benefits [8]. Furthermore, adding titanium dioxide with different weight ratios and particle sizes could improve the reflectivity of coating [9,10]. Hollow glass microspheres were substituted for fillers inside the asphalt concrete, optimizing the proportion for the best cooling effect with a simulated solar heating test [11]. This replacement could improve concrete's emissivity and also reduce thermal conductivity [12]. The lower thermal conductivity materials painted on the asphalt concrete were capable of preventing heat from passing through the coating [13]. The conductive multi-layer structures contributed to diverse applications, such as a radiation-cooling pavement in summer and heat preservation on permafrost [14,15].

In terms of building envelopes, inorganic coating made from mineral powders has been painted on steel plates and concrete specimens. The temperatures and heat fluxes of the coated and uncoated surfaces and interiors were measured in the laboratory and in field tests and compared. The results showed that this coating had an obvious radiation cooling effect and great insulation ability owing to the emissivity and reflectivity. Furthermore, the adhesion and weather resistance were maintained at a certain level [16].

Apart from thermal properties, the mechanical properties of pavements need to be examined as well for practical use. Basic oxygen furnace slag (BOFS) is a byproduct of the steelmaking process. The converter is a by-product of the steelmaking process. A proper ratio for partially replacing aggregate with BOFS should be favorable for the development of bearing capacity and tensile strength when preparing asphalt or recycled concrete [17–21]. Furthermore, based on a two-year on-site test, it was found that the substitution of asphalt concrete entailed a longer life span with less damage in comparison with the original one [22]. Due to the roughness, multi-angularity, and rigidity of BOFS, the substitution concrete had better skid resistance, binding with bitumen, and also abrasion resistance [23]. Even when adopting the warm-mix or hot-mix asphalt approaches, this concrete still showed a great capability for resistance to deflection [24].

In this study, thermal and mechanical properties tests were done on the asphalt concrete specimens, in which aggregates were partially replaced by BOFS. Our previous research found that BOFS has a high far-infrared emissivity. After absorbing solar radiation, these specimens were capable of emitting the stored energy across the sky window (8–13 μm) to the upper sky, reducing energy accumulation and retention in the atmosphere, thereby achieving the goal of urban heat island (UHI) mitigation. The surface skid resistance is needed for the development of a radiation-cooling pavement. Ceramic particles or fine sand particles were added to increase the pavement roughness and the British pendulum number (BPN). BOFS is a byproduct of steelmaking that can not only achieve the goal of radiation cooling but also meet the needs of resource sustainability.

2. Materials and Their Properties

BOFS, a byproduct of steel industries, has lower thermal conductivity compared to natural aggregates. Under the same heat source and with an identical time duration, BOFS has a slower temperature-raising rate. Due to the greater hardness, BOFS applied to asphalt concrete has a bearing capacity comparable to natural aggregates.

In this study, five specimens were designed and prepared, including one general asphalt concrete specimen, and the stone aggregates of the asphalt concrete specimens were replaced by BOFS at proportions of 45 wt.%, 55 wt.%, 65 wt.%, and 75 wt.% respectively. The compositions of the aggregates used in the asphalt concrete specimens are shown in Table 1, and they are dense-graded and conform to ASTM D3515 [25]. The BOFS was also sieved with different sieve sizes. Then, we used BOFS aggregates to replace stone aggregates; the replacement percentage of every sieve size was the same. The content of asphalt to the aggregate mixture was 5.6 wt.%, meeting the specification of between 2 wt.% to 10 wt.%. The naming rules and descriptions of the specimens are shown in Table 2.

Table 1. Passing percentages of aggregates in dense-graded asphalt concrete specimens.

Sieve No.	Sieve Size (mm)	Passing Weight Percentage (%)		
		ASTM Specification	Specimens	Permissible Error
3/8	9.5	90–100	100	±7%
No. 4	4.75	55–85	76	
No. 8	2.36	32–67	46	±6%
No. 16	1.18	-	31	
No. 30	0.60	-	21	±5%
No. 50	0.30	7–23	14	
No. 100	0.15	-	9	±4%
No. 200	0.075	2–10	5.6	±3%

Table 2. Naming and description of the specimens.

Specimen	Description
Benchmark	Standard asphalt concrete
BOF-45	BOFS replacing 45 wt.% natural aggregates
BOF-55	BOFS replacing 55 wt.% natural aggregates
BOF-65	BOFS replacing 65 wt.% natural aggregates
BOF-75	BOFS replacing 75 wt.% natural aggregates

2.1. Emissivity

To identify the radiation cooling effect, the specimens (10 cm in diameter and less than 2 cm in thickness) were placed in the customized apparatus and then heated to 35 °C and 80 °C, respectively. To reflect the reality of usage, it was necessary to confirm that no aggregate was exposed on the measured surface. The environmental causes were reduced as much as possible in the measurements. When the specimens were heated, the infrared light emitted from the specimens was measured by Fourier transform infrared spectrometer (FT-IR) (Invenior, Bruker, Karlsruhe, Germany) in the range of 4000–600 cm^{-1} (3–16.67 μm). The raw data were the result of optical interference.

Through the Fourier transform, the emissivity of heated specimens was calculated compared to the standard blackbody. The emissivity of five specimens was classified into 3–16.7 μm and 8–13 μm, as shown in Figure 1a,b, respectively. The corresponding average emissivity for 3–16.67 μm and 8–13 μm is presented in Table 3. When the substitution of BOFS exceeded 65 wt.%, the emissivity was higher than the benchmark specimen. Among these, the BOF-75 specimen possessed the highest emissivity of 0.88 for 3–16.67 μm and 0.86 for 8–13 μm.

Table 3. Average emissivity of benchmark and BOFS specimens for 3–16.67 μm and 8–13 μm.

Wavelength (μm)	Benchmark	BOF-45	BOF-55	BOF-65	BOF-75
3–16.67	0.844	0.792	0.822	0.872	0.877
8–13	0.829	0.797	0.805	0.870	0.855

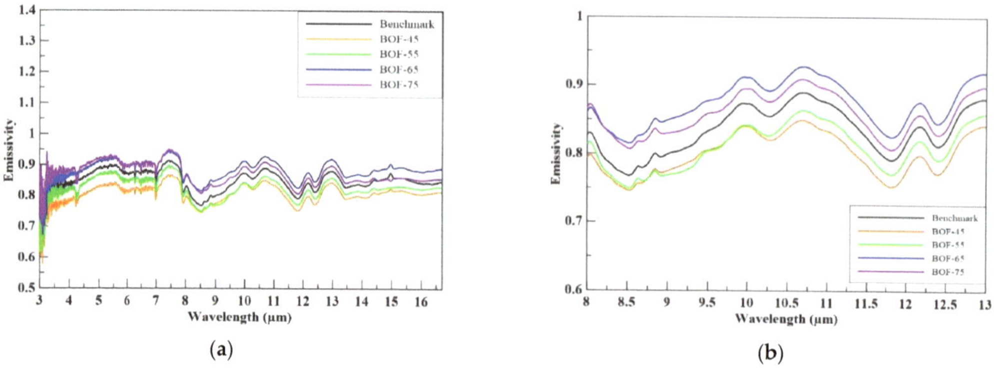

Figure 1. Emissivity of benchmark and basic oxygen furnace slag (BOFS) specimens for (**a**) 3–16.67 μm and (**b**) 8–13 μm.

2.2. Thermal Conductivity

Thermal conductivity is one of the important parameters of thermal performance for materials. The conductivity was tested with a quick thermal conductivity meter (QTM-500, KEM, Tokyo, Japan) used on the specimens. After calibration, the slope equaling the temperature divided by the logarithmic time represents the thermal conductivity coefficient. The larger the slope, the faster the materials conduct heat, and vice versa. In this study, the thermal conductivity of each specimen was averaged.

As illustrated in Figure 2, the thermal conductivity showed a downward trend with the increasing replacement amount of BOFS. Among the specimens, the BOF-75 specimen had the lowest conductivity of 1.17 W/m-K, which means it had better heat-insulating properties than the others. As a consequence, BOFS was taken as a good heat-insulating material.

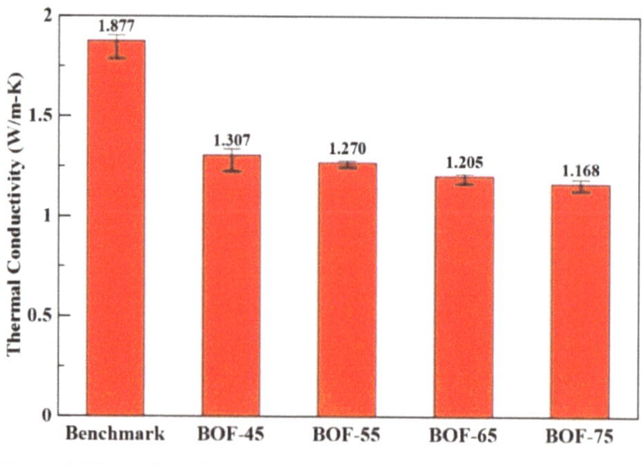

Figure 2. Thermal conductivities of BOFS and benchmark specimens.

3. Experimental Measurement Methods

In this study, the experimental measurement was divided into two aspects: thermal performance and mechanical performance. The thermal performance was measured in the laboratory and also outdoors; the mechanical performance, including the stability value, indirect tensile strength, and British pendulum number, reflects the feasibility of practical application in engineering.

3.1. Indoor Temperature Measurement

To find the best radiation cooling among the benchmark and BOFS specimens, this experimental measurement was conducted in the laboratory and outdoors. The fabrication of the specimen started with a stir-fry pan and was evenly mixed, and the asphalt concrete was poured into the customized steel mold and then compacted to form a specimen. To simulate the actual pavement, the size of the specimen was molded as 50 × 50 cm and with a thickness of 5 cm. Meanwhile, the aggregate layer was 13 cm in thickness underneath the asphalt concrete.

To simulate a sunlight environment in the laboratory, halogen lamps and infrared lamps were used as the heat source in this apparatus to irradiate visible light and infrared light, which provide the majority of solar radiation. The ratios of visible light and infrared light remained at 44% and 53%. Two different radiation intensities, 623 W/m^2 and 436 W/m^2, were tested. According to the previous results of outdoor measurements in summer at Taipei City, the radiation intensity of the heat source should be adjusted to 623 W/m^2, composed of 279 W/m^2 from the halogen lamps erected at 30 cm above the specimen and 344 W/m^2 from the infrared lamp erected at 33 cm above the specimen. To simulate the surface temperature of pavement in winter at Taipei City, the radiation intensity was adjusted to 436 W/m^2 with the maximum surface temperature of 65 °C. Schematic diagrams of the apparatus used in the indoor temperature measurement under the radiation intensities of (a) 623 W/m^2 and (b) 436 W/m^2 are shown in Figure 3.

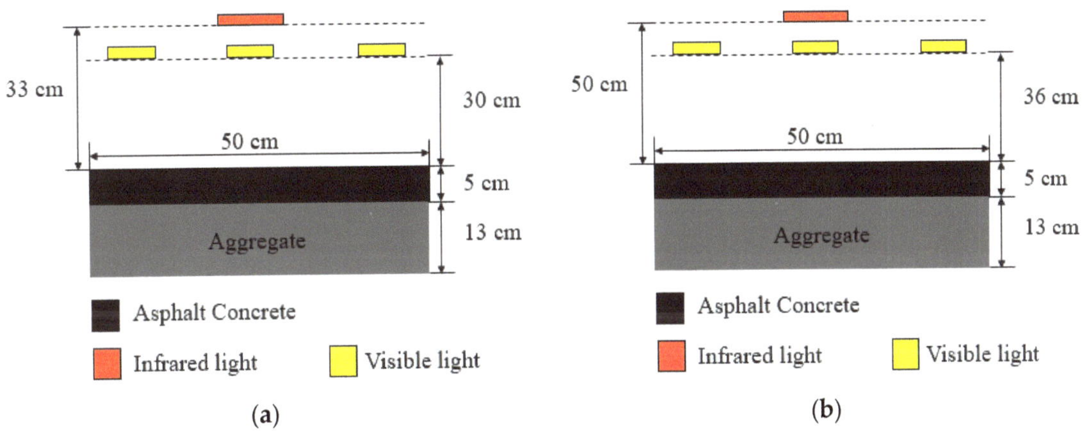

Figure 3. Schematic diagrams of the apparatus used in the indoor temperature measurement under the radiation intensities of (**a**) 623 W/m^2 and (**b**) 436 W/m^2.

The test was conducted in a shaded and airtight room; the specimen was surrounded by Styrofoam. To avoid errors from environmental factors, such as light and ambient temperature, strict conditions were necessary. The thermal cables, which were attached on the surface or buried inside the asphalt concrete, recorded the temperature change at different depths. The experimental apparatus and a schematic diagram of the sensor installation are shown in Figure 4a,b, respectively.

3.2. Outdoor Temperature Measurement

In order to measure their thermal performance, asphalt concrete specimens were exposed to solar radiation, and the outdoor test was conducted on the empty top floor, without shelter, in the summer in 2020 on the roof of the Department of Civil Engineering building, National Taipei University of Technology, Taipei City. The size of the specimens, 50 × 50 × 5 cm, was the same as for those used in the indoor test. Likewise, the aggregate layer was laid 13 cm in thickness beneath the specimens to simulate a real pavement. Also, thermal cables were installed to measure the temperature change, as shown in Figure 4b. The outdoor temperature measurement setup is shown in Figure 5.

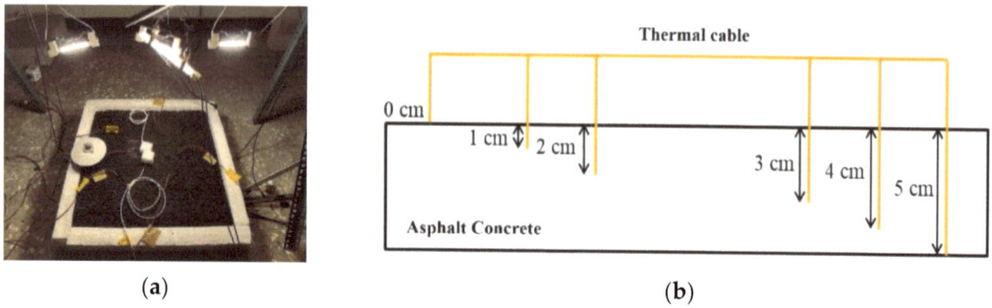

Figure 4. (**a**) Experimental apparatus, (**b**) schematic diagram of sensor installation.

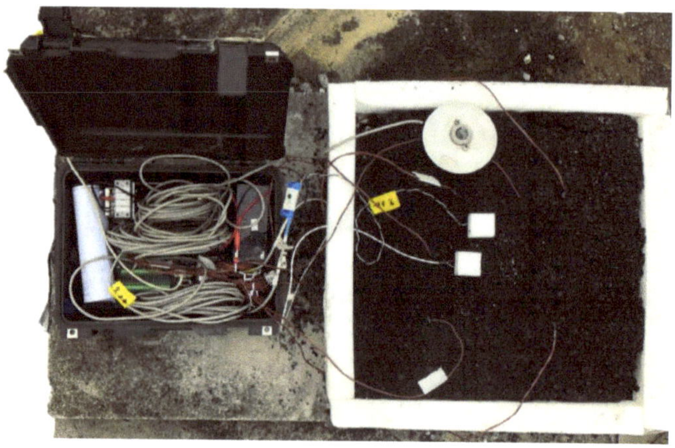

Figure 5. The outdoor temperature measurement equipment and setup.

3.3. Mechanical Properties

The asphalt concrete specimen without BOFS, named the benchmark specimen, and the asphalt concrete with BOFS in the three ratios were prepared with a Marshall apparatus according to the ASTM D6926 standard [26,27]. First, following the mixing formula and substitution ratio, the asphalt and aggregates were mixed by stirring until uniform. The mixture was quickly poured into the steel mold. Then, through 75 times compaction on both sides and then cooling down for about one day, the asphalt concrete specimen was demolded. Its average diameter and height were measured with an electronic Vernier caliper, with an accuracy of 0.01 mm.

The stability value, indirect tensile strength, and British pendulum number are discussed below. Following AI SS-1 set by the Asphalt Institute, the stability values of specimens were measured and specimens were sunk into a constant 60 °C water tank for 30 min and tested with the computer-controlled automatic Marshall apparatus (Ye-Chance Enterprise Co., Taipei, Taiwan) at a uniform rate of 50.8 mm/min. As the pavement is subjected to severe traffic loadings, the stability value must be higher than 8.006 kN to comply with the standard.

The indirect tensile strength was measured according to ASTM D6931 [28]. The specimen was attached between two load stripes and was loaded radially at a speed of 50 ± 5 mm/min. The width of stripes was 12.7 ± 0.3 mm, complying with the standard.

The specimen was exposed to 25 °C water for 30–120 min, and the maximum load at fracture was measured. The indirect tensile strength can be calculated from Equation (1), as follows.

$$St = \frac{2000 \times P}{\pi \times t \times D} \tag{1}$$

In Equation (1), S_t is the indirect tensile strength (kPa), P is the maximum load (N), t is the average thickness of the specimen before test (mm), and D is the average diameter of the specimen before testing (mm). Conforming to ASTM E303, the British pendulum number was measured on the surface of the 50 × 50 × 5 cm specimen with a calibrated British pendulum anti-sliding tester (EL42-6000, ELE International, Leighton Buzzard, UK). In the end, the averaged value was regarded as the BPN value, which represents the anti-skid ability.

4. Results and Discussions

4.1. Indoor Temperature Measurement under Radiation Intensity of 623 W/m^2

The indoor temperature measurement of the specimens was divided into a heating period under a radiation intensity of 623 W/m^2 and cooling period without radiation intensity. The temperature-time relationships and temperature profiles of specimens were drawn and discussed.

4.1.1. Heating Period

The specimens were heated constantly for 24 h under the radiation intensity of 623 W/m^2. Figure 6 shows the temperature–time relationships in the heating period under the radiation intensity of 623 W/m^2. As seen from Figure 6, the specimens reached thermal equilibrium at 24 h. In Figure 6a, the highest temperature of 85 °C occurred in BOF-75 at the end of 24 heating hours and the temperature decreased in the order of BOF-55 and then BOF-45, while the benchmark was 70 °C. No obvious order was observed at the depths of 1 cm and 2 cm, as shown in Figure 6b,c. As seen in Figure 6d, BOF-75 possessed the lowest temperature at the depth of 3 cm.

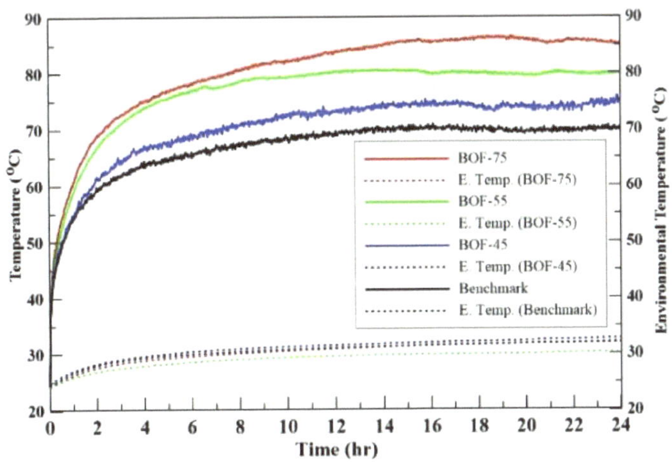

(a)

Figure 6. *Cont.*

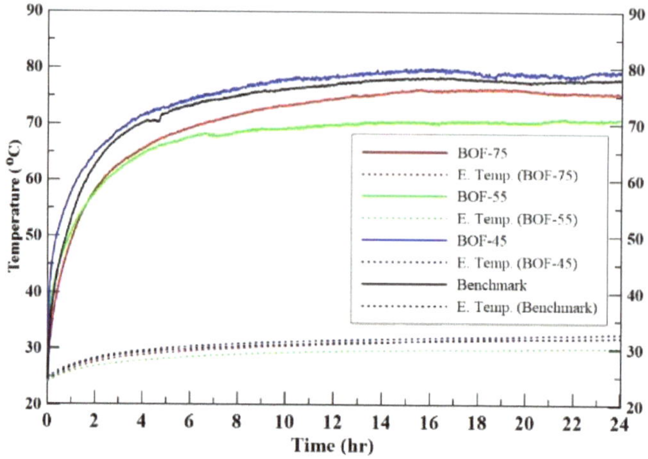

(**b**)

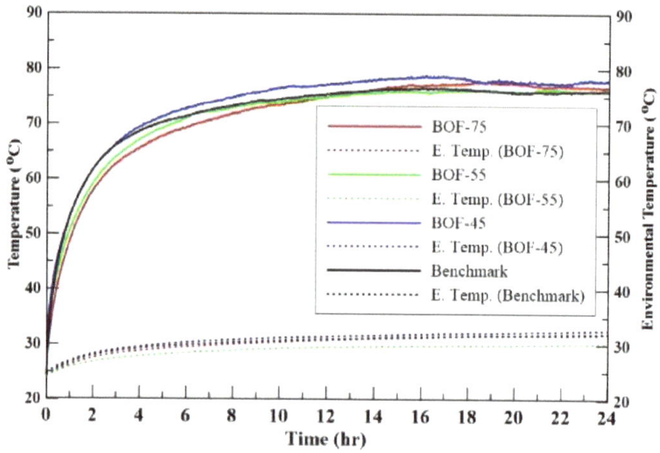

(**c**)

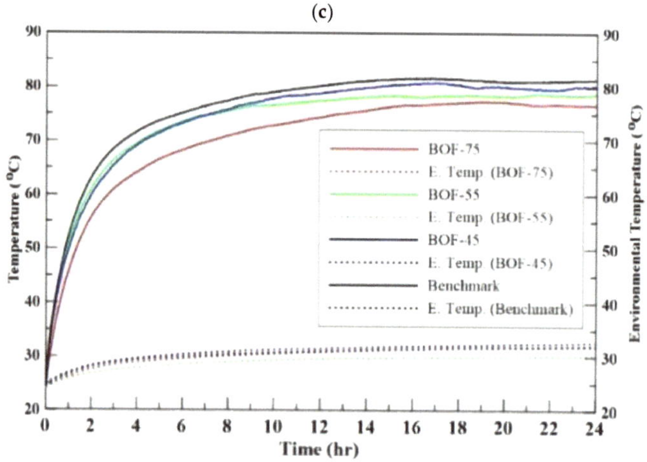

(**d**)

Figure 6. *Cont.*

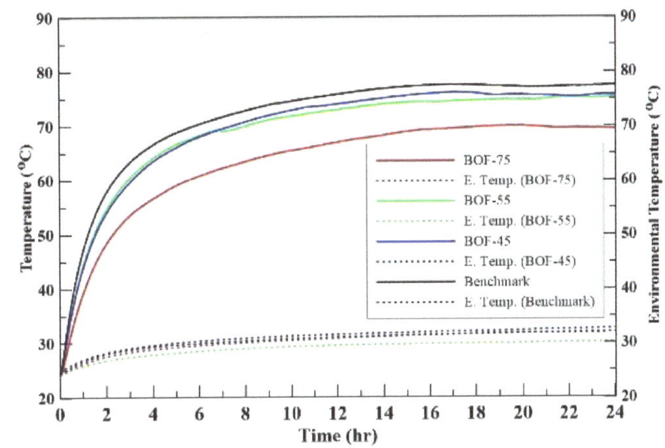

(e)

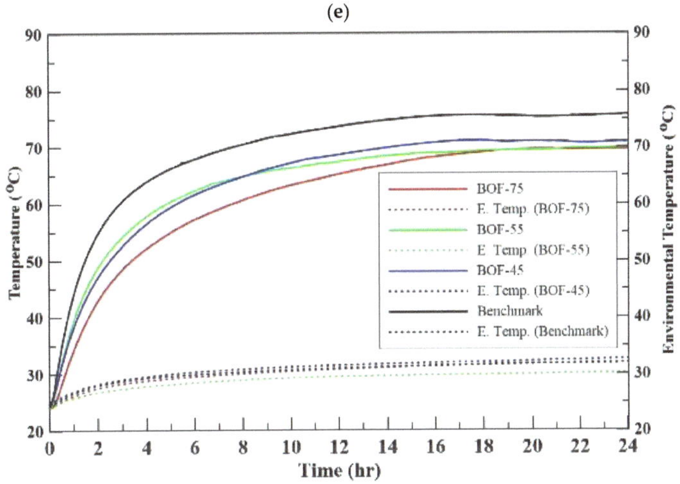

(f)

Figure 6. Temperature–time relationships in the heating period under the radiation intensity of 623 W/m^2 at (**a**) the top surface; (**b**) 1 cm depth; (**c**) 2 cm depth; (**d**) 3 cm depth; (**e**) 4 cm depth; and (**f**) 5 cm depth.

There were two heat sources: one was the halogen and infrared lamps, and the other was the accumulated heat underneath the asphalt concrete; both substantially affected the temperature at the depths of 1 cm and 2 cm.

The order of the temperature on the surface was totally in reverse compared to the temperature at the depths of 3 cm and deeper. The temperatures in profile taken from the thermal equilibrium condition at the end of 24 heating hours are shown in Table 4 and Figure 7. The depth–temperature curves of BOF-55 and BOF-75 were similar. In the profile, their temperatures on the surface were higher than inside and broadly decreased when going deeper. On the other hand, the depth–temperature curve of the benchmark and BOF-45 were similar in another way: the highest temperature was recorded at the depth of 3 cm. Compared to the benchmark specimen, although all the specimens with BOFS replacements had hotter surfaces, their temperatures were lower starting at 3 cm depths, especially BOF-75.

Table 4. Temperature (°C) corresponding to different depths at the end of 24 heating hours under the radiation intensity of 623 W/m^2.

Depth \ Specimen	Benchmark	BOF-45	BOF-55	BOF-75
0 cm	69.86	75.14	79.88	85.10
1 cm	77.94	79.56	70.83	75.33
2 cm	76.16	78.06	75.94	76.72
3 cm	81.40	80.20	78.71	76.63
4 cm	77.49	75.75	75.29	69.42
5 cm	75.77	70.98	69.97	69.64

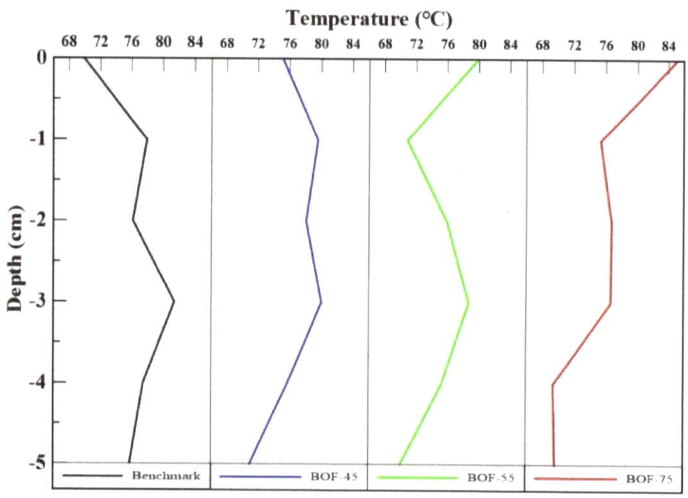

Figure 7. Depth–temperature profile of specimens at the end of 24 heating hours under the radiation intensity of 623 W/m^2.

4.1.2. Cooling Period

All the lamps were turned off after 24 h of heating; the temperature change in the subsequent 1-h cooling was recorded and shown in Figure 8. When the heat source was removed, the heat obviously dissipated from the specimen. The cooling rate at each depth gradually became similar, and the temperature–time relationships of all specimens at the depth of 3 cm were the most concentrated.

The depth–temperature profiles of specimens within five cooling hours in the cooling period are shown in Figure 9. In the beginning of the cooling period, the sequences of the surface temperatures are the same. Notably, all the specimens containing BOFS possessed a higher temperature at a depth of 5 cm than 3 cm. As seen from Figure 9, the temperature of the benchmark specimen at the depth of 3 cm was higher than the temperature at 5 cm in the first hour, but the temperature at a depth of 3 cm was less than the temperature at a depth of 5 cm at other times.

4.2. Indoor Temperature Measurement under Radiation Intensity of 436 W/m^2

The indoor temperature measurement of the specimens was divided into a heating period under a radiation intensity of 436 W/m^2 and a cooling period without radiation intensity. The temperature–time relationships and temperature profiles of specimens were determined and discussed.

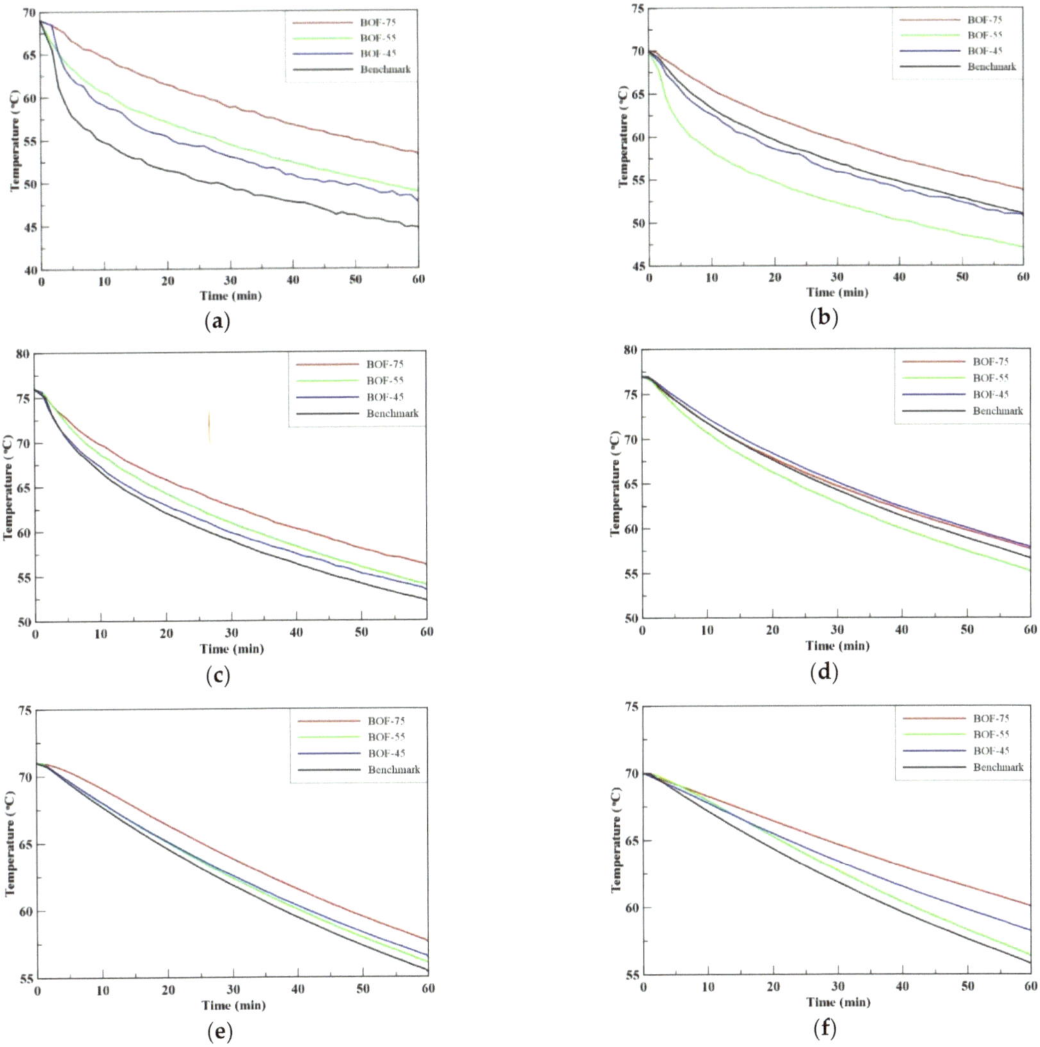

Figure 8. Temperature–time relationships in the cooling period at (**a**) the top surface; (**b**) 1 cm depth; (**c**) 2 cm depth; (**d**) 3 cm depth; (**e**) 4 cm depth; and (**f**) 5 cm depth.

4.2.1. Heating Period

The specimens were heated constantly for 24 h under the radiation intensity of 436 W/m^2 until the temperatures of the specimens reached a stable state. As shown in Figure 10a, BOF-75 had the highest temperature of 66.82 °C and BOF-55 had the next highest temperature of 64.74 °C, corresponding with the results under the radiation intensity of 623 W/m^2. Although the order between BOF-45 and the benchmark was disrupted and did not correspond with the previous results, the temperature difference between 56.95 °C and 57.69 °C was not significant. As shown in Figure 10b,c, the temperature changes at the depths of 1 cm and 2 cm were similar to the results under the radiation intensity of 623 W/m^2. Influenced by the upper and lower heat sources, it was difficult to find the order at these two depths. As shown in Figure 10d, the temperatures at the depth of 3 cm were in descending order of BOF-75 (63.46 °C), BOF-55 (63.53 °C), BOF-45

(64.17 °C), and then the benchmark (65.65 °C). This order happened for the results under the radiation intensity of 623 W/m^2, which was in reverse on the surface. As shown in Figure 10e,f, the temperature–time relationship at the depths of 4 cm and 5 cm were similar to those in Figure 10d.

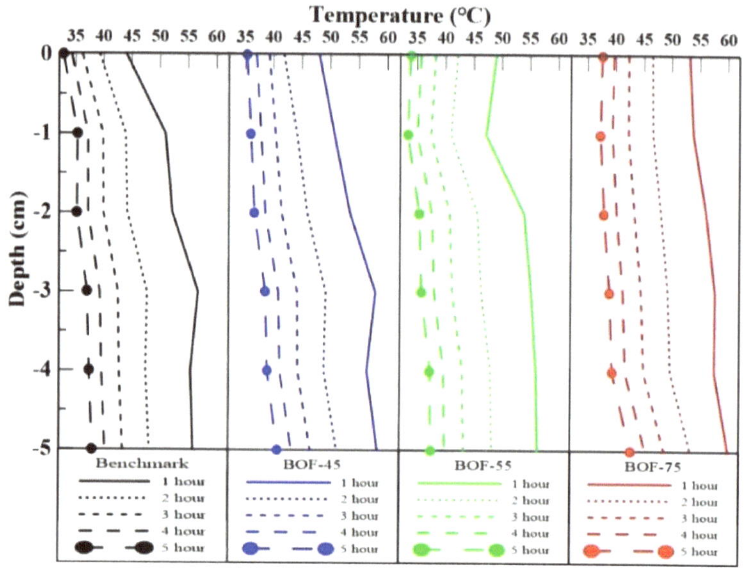

Figure 9. Depth–temperature profile of specimens within five cooling hours in the cooling period.

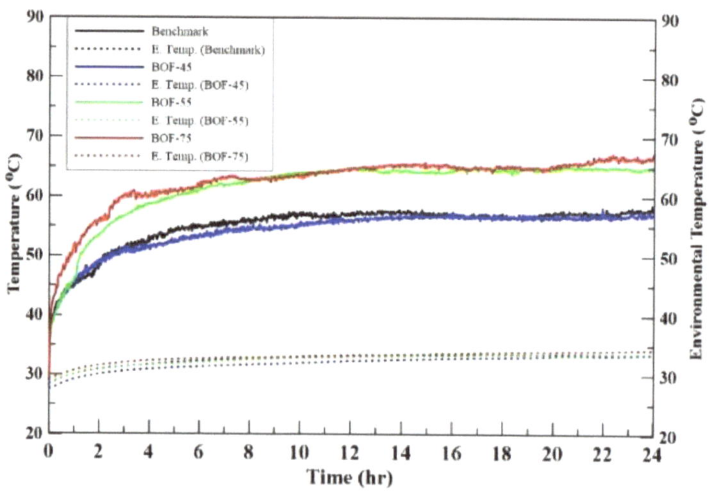

(**a**)

Figure 10. *Cont.*

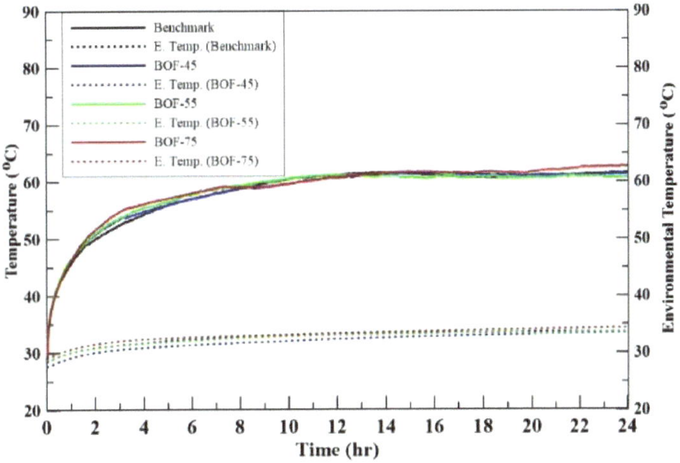

(**b**)

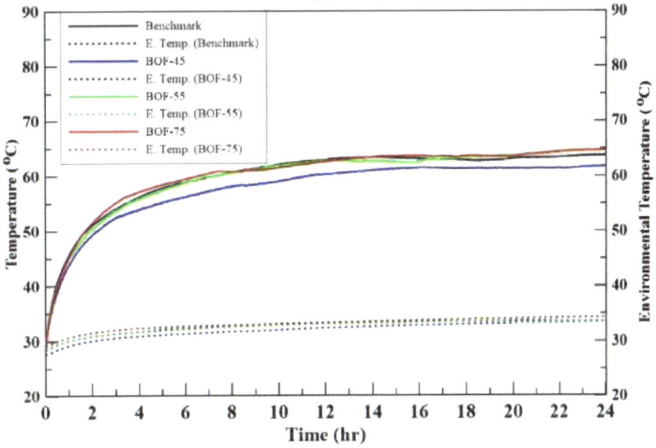

(**c**)

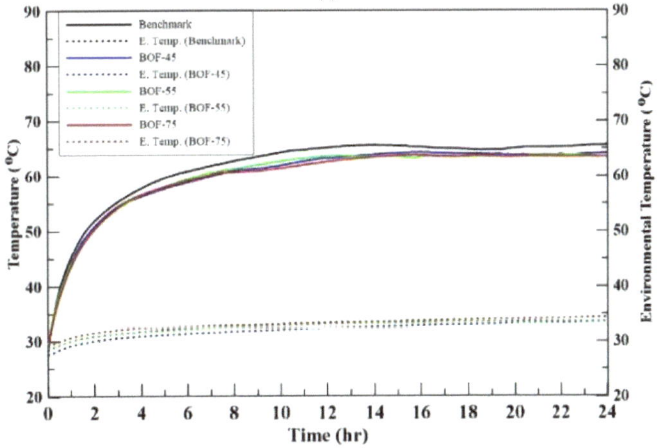

(**d**)

Figure 10. *Cont.*

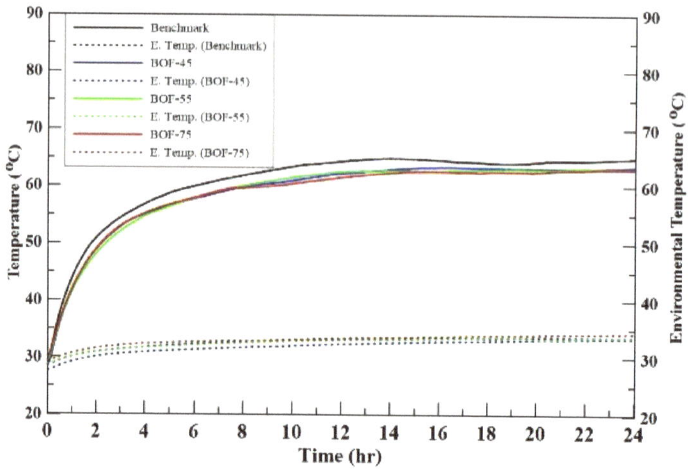

(e)

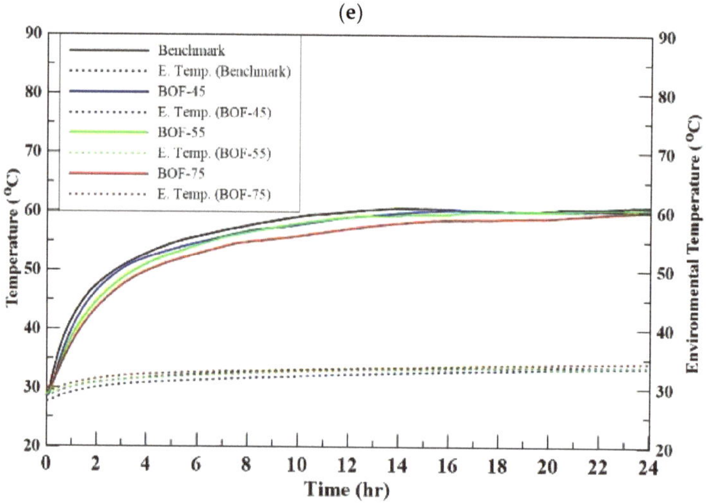

(f)

Figure 10. Temperature–time relationships in the heating period under the radiation intensity of 436 W/m^2 at (**a**) the top surface; (**b**) 1 cm depth; (**c**) 2 cm depth; (**d**) 3 cm depth; (**e**) 4 cm depth; and (**f**) 5 cm depth.

The depth–temperature profile taken from the balanced condition at the end of 24 heating hours is shown in Table 5 and Figure 11. In the profile, the temperatures of all specimens declined constantly from 3 cm to 5 cm depths. Furthermore, it clearly shows that BOF-75 possessed the lowest temperature of 63.07 °C at the depth of 4 cm and 60.04 °C at the depth of 5 cm. Furthermore, the depth–temperature curves of BOF-55 and BOF-75 were similar. In the profile, their temperatures at the depth of 3 cm were higher than at the depth of 5 cm, and the temperatures on the surface were higher than that at the depth of 3 cm. On the other hand, the depth–temperature curves of the benchmark and BOF-45 were similar in a different way. These two specimens had the highest temperatures at the depth of 3 cm and the lowest temperatures on the surface. All the trends above were similar to the results under the radiation intensity of 623 W/m^2.

Table 5. Temperature (°C) corresponding to different depths at the end of 24 heating hours under the radiation intensity of 436 W/m^2.

Depth \ Specimen	Benchmark	BOF-45	BOF-55	BOF-75
0 cm	57.69	56.95	64.74	66.82
1 cm	61.63	61.43	60.96	62.83
2 cm	63.78	61.76	64.60	64.69
3 cm	65.65	64.17	63.53	63.46
4 cm	64.97	63.49	61.15	63.07
5 cm	60.88	60.40	60.28	60.04

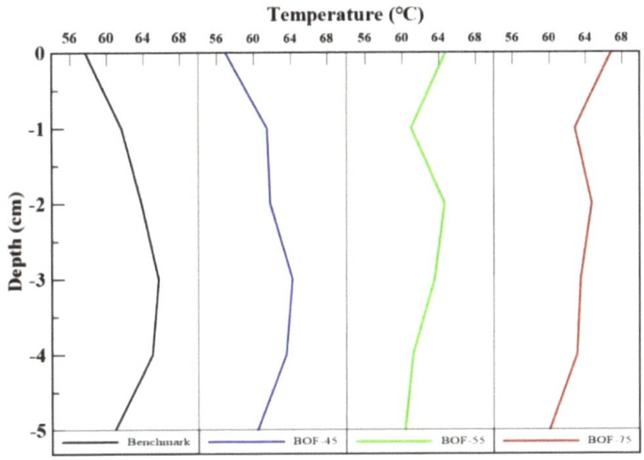

Figure 11. Depth–temperature profile of specimens at the end of 24 heating hours under the radiation intensity of 436 W/m^2.

4.2.2. Cooling Period

After 24-h heating, the temperature change in the subsequent 1-h cooling was recorded, as shown in Figure 12. When the heat source was removed, the heat obviously dissipated from the specimens, especially those which possessed higher conductivity. BOF-75 had the lowest thermal conductivity, leading to the slowest cooling rate. In contrast, the benchmark specimen had the highest thermal conductivity, leading to the fastest cooling rate. Also, the cooling rates for each depth gradually became similar, and the rates at the depth of 3 cm were the most concentrated.

The depth–temperature profiles of specimens within five cooling hours can be seen in Figure 13. The temperature difference between 3 cm and 5 cm depths deserves discussion. At the end of five hours, BOF-75 had the lowest temperature on the surface, and its temperature at the depth of 3 cm was lower than that at 5 cm. With regard to BOF-55, its temperature at the depth of 3 cm was lower than that at 5 cm only in the first hour, with a slight difference, and its temperature on the surface was lower than that at the depth of 3 cm all the time. Also, the temperature of the benchmark at the depth of 3 cm was higher than that at 5 cm until the fourth hour.

To sum up, compared to the benchmark asphalt concrete, no matter which radiation intensity was taken, the specimens with partial BOFS replacement showed faster cooling rates when the light was turned off, which might refer to more heat accumulating inside the body.

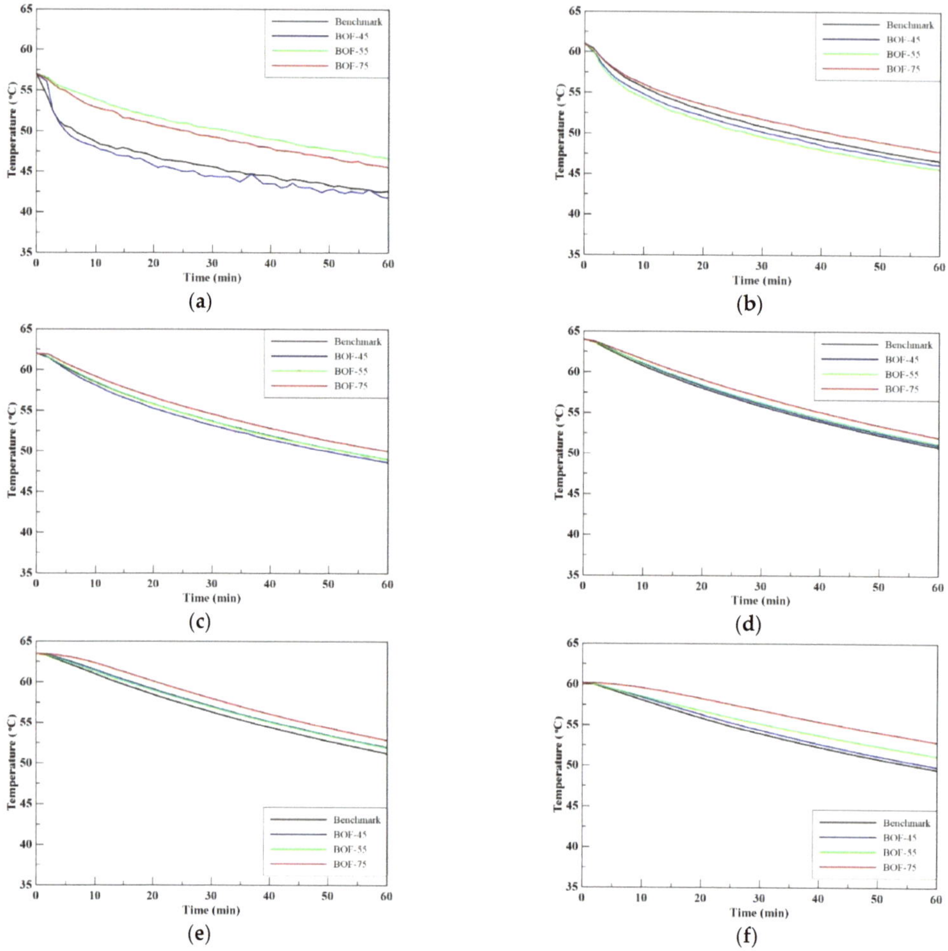

Figure 12. Temperature–time relationships in the cooling period at (**a**) the top surface; (**b**) 1 cm depth; (**c**) 2 cm depth; (**d**) 3 cm depth; (**e**) 4 cm depth; and (**f**) 5 cm depth.

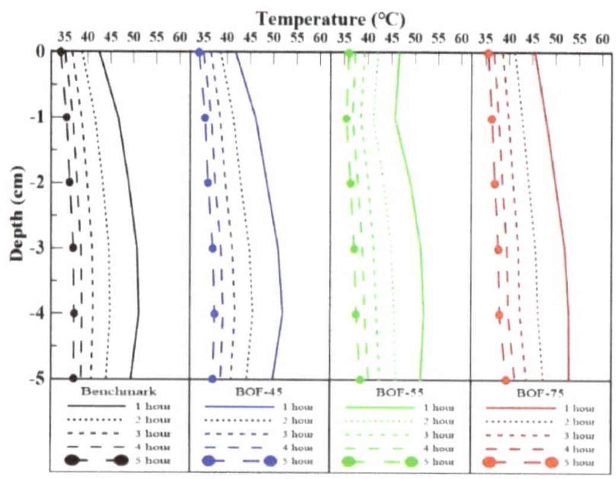

Figure 13. Depth–temperature profile of specimens within five cooling hours in the cooling period.

4.3. Outdoor Temperature Measurement

The temperature–time relationships of specimens in the outdoor test, across both daytime and nighttime for a full 24 h, were measured and are shown in Figure 14. The specimens' temperature and environmental information were recorded and are shown in Table 6. The field test results were easily affected by the environmental factors, especially after sunset. Therefore, it is more reliable to use the average ambient temperature for comparison. As shown in Table 6, during the daytime, the benchmark had a higher temperature than the ambient temperature of about 35 °C. Also, unlike the indoor test, specimens could not obtain long-term and sufficient heat from the sun. As a consequence, for BOF-75, its low thermal conductivity and large heat capacity resulted in relatively low temperatures. The calculation of the heat capacity is explained in Section 4.4.

As shown in Table 6, using the ambient temperature of about 30 °C during nighttime as the comparison basis, the temperatures of the benchmark specimen were generally higher than those of BOF-55 and BOF-75. In terms of results, the use of BOFS did have a positive effect on pavement cooling.

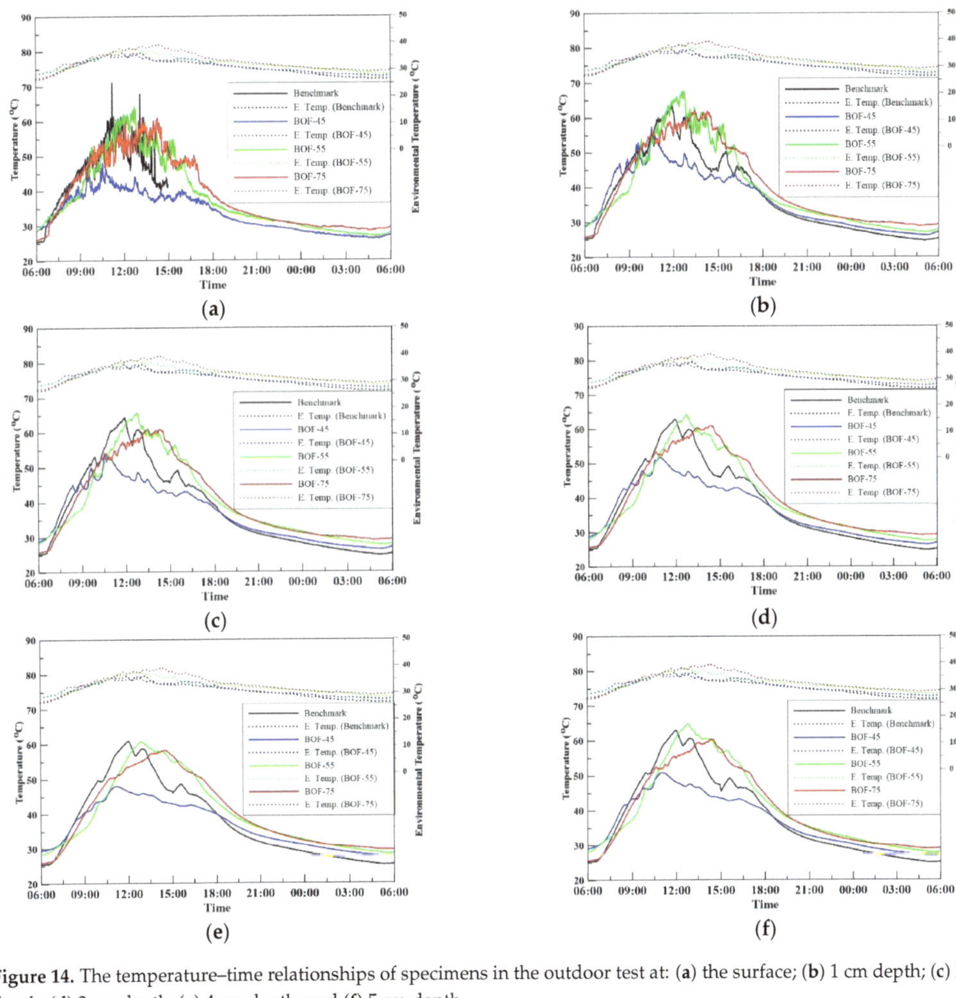

Figure 14. The temperature–time relationships of specimens in the outdoor test at: (**a**) the surface; (**b**) 1 cm depth; (**c**) 2 cm depth; (**d**) 3 cm depth; (**e**) 4 cm depth; and (**f**) 5 cm depth.

Table 6. Environmental information and temperature corresponding to different depths.

Specimen		Benchmark		BOF-45		BOF-55		BOF-75	
Time Period		10:00 ∫ 14:50	20:00 ∫ 22:30	10:00 ∫ 13:10	20:00 ∫ 23:30	9:20 ∫ 11:40	22:40 ∫ 04:10	9:20 ∫ 10:50	22:40 ∫ 06:00
Average Temperature (°C)		35	30	35	30	35	30	35	30
Average Humidity (%)		42.23	70.81	43.90	70.13	38.40	77.26	44.70	74.24
Average Solar Radiation (W/m^2)		642.73	0	412.75	0	756.34	0	646.32	0
Average Temperature (°C) at Various Depths	0 cm	51.04	29.59	41.88	29.84	50.97	29.01	49.63	29.55
	1 cm	54.54	30.32	49.11	30.82	53.09	29.39	52.21	30.01
	2 cm	55.98	30.76	48.93	31.17	51.38	30.36	50.55	30.27
	3 cm	55.67	31.11	48.53	31.47	49.87	30.27	49.64	30.40
	4 cm	54.69	31.61	46.22	32.96	45.99	31.36	46.73	31.05
	5 cm	56.25	31.58	48.45	32.26	49.32	30.80	48.87	30.35

4.4. Heat Capacity, Newton's Cooling and Radiation Cooling

To confirm the radiation cooling effectiveness of specimens, the loss by radiation cooling was quantified based on several assumptions as follows. The heat which enters the asphalt concrete and then transfers to the ground has only a small impact on pedestrian and human activities; the heat below the specimens was not discussed. On the other hand, the heat on the upper surface of the asphalt concrete specimens is worth further study.

Ideally, when objects receive heat, they transfer it in three ways: reflection, transmission, and absorption. The sum of the energy from these ways must be equal to the energy incident, an illustration of which is shown in Figure 15a. However, it is hard to determine the reflection, transmission, and absorption of specimens realistically, even using a pyranometer. It is impossible to know how much heat has dissipated through radiation. We can only assume that, following the law of conversation of energy, the relationship between the energy entering and leaving the specimen is constant. In this study, the focus was on the cooling period, as shown in Figure 15b. As the temperature of the test body cooled over time under the radiation intensity of 623 W/m^2, the time interval between data was set to 5 min.

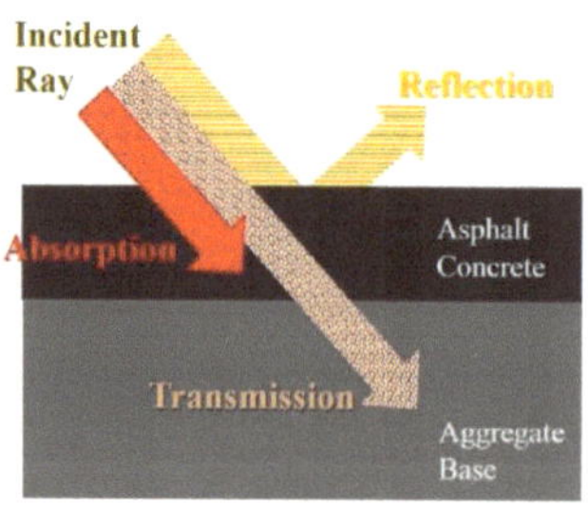

(a)

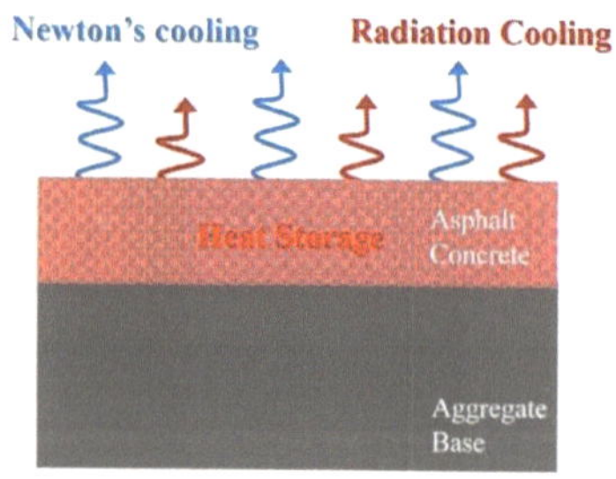

(b)

Figure 15. Relationship between the pavement and (**a**) basic heat transfer and (**b**) radiation cooling.

The heat stored in a specimen at various time intervals can be calculated as in Equation (2). In this function, Q is the heat energy in storage (J), m is the mass of the specimen (kg), c is the heat capacity of the specimen (J/kg-K), and ΔT is the change in temperature (K).

$$Q = mc\Delta T \tag{2}$$

The specimen was divided into five equal parts each with a thickness of 5 cm; the sectional drawing of the specimen is shown in Figure 16. In each part, the temperature at the center was the average of the adjoining temperature at the upper and lower part. Then, the sum of the heat stored in each part at the *t* minute in the cooling period represented the heat storage in the entire specimen at that time. Finally, the heat storage of (t + 5) minutes was subtracted from that at the t minute, and added up over 24 h to obtain the final heat storage of the specimen. This formula is shown in Equation (3), and the calculation results for each hour are presented in Table 7.

$$\text{The final heat storage in 24 h} = \sum_{t=0}^{288} (Q_{5t}) - \left(Q_{5(t+1)}\right) \tag{3}$$

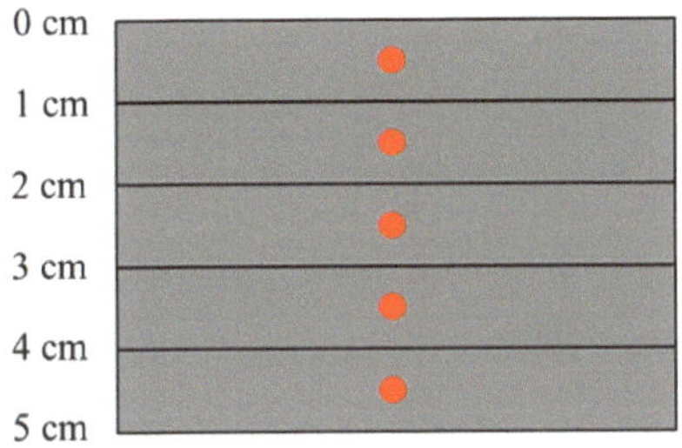

Figure 16. Sectional drawing of the specimen.

Table 7. Heat energy in storage at a different hour (unit: J).

Hour \ Specimen	Benchmark	BOF-45	BOF-55	BOF-75
1	3871.92	3928.44	3956.71	4691.52
2	2232.71	1695.73	2430.55	2656.65
3	1441.37	1526.16	1413.11	1610.94
4	904.39	960.91	1017.44	1187.01
5	678.29	763.08	932.65	932.65
6	536.98	423.93	650.03	734.82
7	423.93	367.41	452.20	508.72
...	...	...	...	...
Sum (24 h)	255,998.94	233,841.38	288,641.77	306,616.52

Based on Newton's law of cooling, the heat stored in the pavement is also transferred into the atmosphere, which is called "Newton's cooling" in this study, as shown in Equation (4):

$$q = hA\Delta T \tag{4}$$

In this function, q is the power of Newton's cooling transferred out of the specimen (watts), h is the heat transfer coefficient (W/m^2-K), A is the heat transfer surface area (m^2), and ΔT is the temperature difference between the environment and specimen (K). The final value for the difference, resulting from Newton's cooling, between specimen and air can be calculated as in Equation (5). Notably, it was necessary to multiply the difference between the total heat of t and that at (t + 5) minutes by a time interval of 5 min. The calculation results for each hour are presented in Table 8.

$$\text{The final Newton's cooling in 24 h} = \sum_{t=0}^{288} \left[(q_{5t}) - \left(q_{5(t+1)}\right) \times 5\right] \tag{5}$$

Table 8. Heat energy for Newton's cooling at different hours (unit: J).

Hour \ Specimen	Benchmark	BOF-45	BOF-55	BOF-75
1	184.69	196.26	292.53	333.16
2	102.92	23.08	171.10	160.02
3	75.39	80.00	83.71	102.77
4	38.81	51.77	61.17	75.05
5	33.77	52.61	48.01	53.64
6	26.46	15.76	29.53	41.18
7	16.92	21.13	22.06	33.01
...	...	...	...	...
Sum (24 h)	14,186.33	13,545.51	22,342.61	23,985.92

Finally, following the above data in Tables 7 and 8, Newton's cooling energy subtracted from the stored heat energy equals the radiation cooling energy, as shown in Table 9.

Table 9. The heat energy in radiation cooling at different hours (unit: J).

Hour \ Specimen	Benchmark	BOF-45	BOF-55	BOF-75
1	3687.23	3732.19	3664.18	4358.36
2	2129.79	1672.65	2259.45	2496.63
3	1365.98	1446.16	1329.40	1508.18
4	865.58	909.14	956.27	1111.96
5	644.52	710.47	884.64	879.01
6	510.52	408.17	620.50	693.64
7	407.01	346.28	430.14	475.71
...	...	...	...	...
Sum (24 h)	241,812.61	220,295.87	266,299.16	282,630.60

The overall amounts of the accumulated heat storage energy, Newton's cooling energy, and the radiation cooling energy are presented in Tables 10–12, respectively. As shown in Table 10, heat storage energy within 7 h accounted for up to 80% of heat storage energy within 24 h. We can deduce that the main cooling was completed within 7 h in the cooling period. Also, BOF-75 had the highest heat storage energy among all the specimens, as shown in Table 10. In the seventh cumulated hour, Newton's cooling energy accounted for about 90% of the stored energy, as shown in Table 11. Therefore, the main way of dissipating heat after 7 h was radiation cooling.

Table 10. Accumulated heat storage energy in the cooling period (unit: J).

Cumulated Hour \ Specimen	Benchmark	BOF-45	BOF-55	BOF-75
1	138,428.22	116,892.42	160,698.82	156,685.59
2	172,201.54	149,337.42	196,591.81	198,061.44
3	191,730.71	167,905.68	217,675.40	222,536.50
4	204,759.58	180,623.67	231,636.93	238,787.26
5	214,029.58	190,091.50	241,500.43	251,024.79
6	220,869.03	197,100.53	249,018.18	260,038.69
7	226,323.64	203,063.85	254,811.92	267,190.77
...	...	...	...	...
24	255,998.94	233,841.38	288,641.77	306,616.52

Table 11. Accumulated Newton's cooling energy in the cooling period (unit: J).

Cumulated Hour \ Specimen	Benchmark	BOF-45	BOF-55	BOF-75
1	9842.91	8253.74	15,271.91	15,796.84
2	11,389.49	9985.62	17,720.81	18,507.62
3	12,221.99	10,897.82	19,061.52	20,011.26
4	12,710.38	11,506.13	19,878.20	20,938.10
5	13,053.25	11,596.67	20,421.84	21,607.52
6	13,275.10	12,249.19	20,813.72	22,065.73
7	13,447.55	12,511.17	21,105.98	22,418.62
...	...	...	...	...
24	14,186.33	13,545.51	22,342.61	23,985.92

Table 12. Accumulated radiation cooling energy in the cooling period (unit: J).

Cumulated Hour \ Specimen	Benchmark	BOF-45	BOF-55	BOF-75
1	128,585.31	108,638.68	145,426.92	140,888.75
2	160,812.05	139,351.81	178,871.00	179,553.83
3	179,508.72	157,007.86	198,613.89	202,525.24
4	192,049.20	169,117.54	211,758.73	217,849.17
5	200,976.34	178,134.83	221,078.60	229,417.27
6	207,593.93	184,851.34	228,204.45	238,002.96
7	212,876.09	190,552.68	233,705.94	244,772.15
...	...	...	...	...
24	241,812.61	220,295.87	266,299.16	282,630.60

In the above calculations, it seems that, under the same radiation intensity, the specimens with higher heat capacities could absorb more heat inside the specimen, and less heat escaped to the environment. And according to Newton's cooling law, the thermal equilibrium occurs under the condition of releasing sufficient energy for the specimen. BOF-75 had a higher surface temperature because it stored as well as absorbed more heat, which is a reasonable result. As far as radiation cooling is concerned, a larger value is considered better, which means that more heat leaves the specimen in this way. BOF-75 had the largest radiation cooling among the specimens.

Based on the results, it can be concluded that BOF-75 had better heat capacity and could dissipate less heat into the environment and absorb more heat than other specimens. Although a higher surface temperature occurred under the same radiation intensity, BOF-75 possessed the best radiation cooling ability in the cooling period among the specimens, and the heat could be transferred into long-wave radiation, which is not easily absorbed by air. In addition, the material emissivity results measured by the Fourier Infrared Spectrometer (FTIR) help to show that BOF-75 demonstrated a better performance in the radiation cooling project.

4.5. Mechanical Test Results

The stability values of the specimens were measured according to ASTM D6927 [27]. As shown in Table 13 and Figure 17, the stability values of all specimens were far larger than the standard requirement of 8.006 kN, which was set by the Asphalt Institute. As the replacement ratio of BOFS increased, the stability value increased. Among all specimens, BOF-75 had the highest average stability value of 34.54 kN.

Table 13. Stability value of the specimens.

Specimen	No.	Stability Value (kN)	Average Stability Value (kN)
Benchmark	1	12.13	11.61
	2	12.25	
	3	10.43	
BOF-45	1	16.98	17.07
	2	17.68	
	3	16.55	
BOF-55	1	17.05	15.91
	2	15.67	
	3	15.02	
BOF-75	1	35.09	34.54
	2	33.59	
	3	34.93	

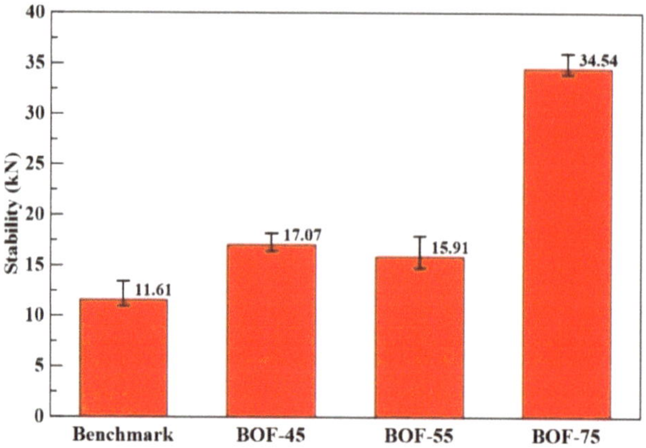

Figure 17. Stability values of the specimens.

Indirect tensile strength, one of the indexes of asphalt concrete, can be inferred from the resistance of rut and crack. Following ASTM D6931 [28], the indirect tensile strength was measured and calculated as shown in Table 14 and Figure 18. It implied that BOFS had a better binding ability with bitumen than natural aggregate did.

Table 14. Indirect tensile strength of the specimens.

Specimen	Maximum Load (kN)	Indirect Tensile Strength (MPa)	Average Indirect Tensile Strength (MPa)
Benchmark	5.592	0.511	0.568
	6.524	0.612	
	6.259	0.580	
BOF-45	11.75	1.255	1.253
	12.09	1.251	
	12.02	1.253	
BOF-55	13.64	1.509	1.414
	12.26	1.364	
	12.30	1.368	
BOF-75	12.54	1.456	1.253
	11.08	1.264	
	9.369	1.038	

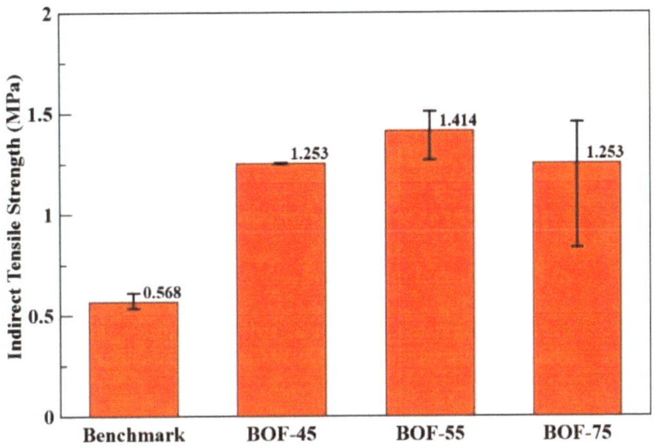

Figure 18. Indirect tensile strength of the specimens.

The BPN was measured according to ASTM E303 [29]. As shown in Table 15, all specimens conformed to the BPN of 45 [30]. Specimens replaced partially by BOFS demonstrated an anti-skid ability, while the flatness was roughly the same. There is no doubt that using asphalt concrete pavement in which natural aggregates have been partially replaced by BOFS affects driving safety.

Table 15. British pendulum number (BPN) of the specimens.

Specimen No.	1	2	3	4	Average
Benchmark	82	83	84	83	83
BOF-45	80	80	81	80	80
BOF-55	79	79	79	80	79
BOF-75	82	80	80	82	81

5. Conclusions

Regarding the asphalt concrete with partial BOFS replacement in different proportions, several conclusions can be stated as follows.

1. The BOF-75 specimen had a high emissivity of 0.86 across the sky window (8–13 μm) and the thermal conductivity of the specimen decreased as the substitution ratio of BOFS increased.
2. Under two different radiation intensities, the surface temperature order among the specimens was the same. Furthermore, the temperatures from high to low were in the order BOF-75 > BOF-55 > BOF-45 > benchmark specimen.
3. At the end of 24 heating hours, the temperature order at the depth of 3 cm was opposite to that for the surface, and the temperatures at the depth of 3 cm, from high to low, were in the order benchmark asphalt concrete > BOF-45 > BOF-55 > BOF-75. When the depth was more than or equal to 3 cm, the temperatures remained in the same order.
4. According to the quantitative calculation of the heat storage, under the same radiation intensity, the BOF-75 specimen absorbed the most heat inside the body, contributing to less heat remaining in the environment. In the cooling period, Newton's cooling energy accounted for about 90% of the stored energy within 7 h, so the main ways of heat dissipation after the seventh hour were radiation and conduction cooling, corresponding to the emission across the urban boundary layer. Therefore, the BOF-75 specimen was considered more favorable to urban heat island mitigation.

5. The anti-skid index, the BPN value, was far larger than the standard of 45; the stability value complied with the standard of 8.006 kN, and BOF-75 could reach 34.54 kN. Consequently, BOFS can be successfully applied in pavements.
6. All asphalt concrete with partial BOFS replacement had a greater indirect tensile strength than the traditional concrete. It is very likely that BOFS has a better binding ability with bitumen than natural aggregate does.
7. From the test results, it can be seen that BOFS has great applicability in pavements due to its thermal performance and mechanical properties.

Author Contributions: Conceptualization, Y.-F.L. and C.-H.H.; Data curation P.-A.Y.; Formal analysis, P.-A.Y. Investigation, P.-A.Y. and C.-H.W.; Methodology, Y.-F.L. and C.-H.H.; Project administration, Y.-F.L.; Supervision, Y.-F.L. and T.-W.C.; Writing—original draft, P.-A.Y. and C.-H.W.; Writing—review & editing, Y.-F.L. and T.-W.C. All authors have read and agreed to the published version of the manuscript.

Funding: This research was funded by the Ministry of Education of Taiwan, under contract No. L7081101-4, and the "Research Center of Energy Conservation for New Generation of Residential, Commercial, and Industrial Sectors" from the Ministry of Education of Taiwan, under contract No. L7091101-19.

Institutional Review Board Statement: Not applicable.

Informed Consent Statement: Not applicable.

Data Availability Statement: Not applicable.

Conflicts of Interest: The authors declare no conflict of interest.

References

1. NOAA Climate, Climate Change: Global Temperature. Available online: https://www.climate.gov/news-features/understanding-climate/climate-change-global-temperature (accessed on 22 July 2020).
2. Cao, X.; Tang, B.; Zhu, H.; Zhang, A.; Chen, S. Cooling principle analyses and performance evaluation of heat-reflective coating for asphalt pavement. *J. Mater. Civ. Eng.* **2011**, *23*, 1067–1075. [CrossRef]
3. Chen, Q.; Wang, C.; Fu, H.; Zhang, L. Durability evaluation of road cooling coating. *Constr. Build. Mater.* **2018**, *190*, 13–23. [CrossRef]
4. Synnefa, A.; Karlessi, T.; Gaitani, N.; Santamouris, M.; Assimakopoulos, D.N.; Papakatsikas, C. Experimental testing of cool colored thin layer asphalt and estimation of its potential to improve the urban microclimate. *Build Environ.* **2011**, *46*, 38–44. [CrossRef]
5. Anting, N.; Din, M.F.M.; Iwao, K.; Ponraj, M.; Siang, A.J.L.M.; Yong, L.Y.; Prasetijo, J. Optimizing of near infrared region reflectance of mix-waste tile aggregate as coating material for cool pavement with surface temperature measurement. *Energy Build.* **2011**, *158*, 172–180. [CrossRef]
6. Zheng, M.; Han, L.; Wang, F.; Mi, H.; Li, Y.; He, L. Comparison and analysis on heat reflective coating for asphalt pavement based on cooling effect and anti-skid performance. *Constr. Build. Mater.* **2015**, *93*, 1197–1205. [CrossRef]
7. Xie, N.; Li, H.; Abdelhady, A.; Harvey, J. Laboratorial investigation on optical and thermal properties of cool pavement nano-coatings for urban heat island mitigation. *Build Environ.* **2019**, *147*, 231–240. [CrossRef]
8. Jiang, L.; Wang, L.; Wang, S. A novel solar reflective coating with functional gradient multilayer structure for cooling asphalt pavements. *Constr. Build. Mater.* **2019**, *210*, 13–21. [CrossRef]
9. Sha, A.; Liu, Z.; Tang, K.; Li, P. Solar heating reflective coating layer (SHRCL) to cool the asphalt pavement surface. *Constr. Build. Mater.* **2017**, *139*, 355–364. [CrossRef]
10. Huang, Z.; Ruan, X. Nanoparticle embedded double-layer coating for daytime radiation cooling. *Int. J. Heat Mass Transf.* **2017**, *104*, 890–896. [CrossRef]
11. Feng, C.; Zhang, H.; Li, C.; Jia, W.; Lai, F. The effects of hollow glass microsphere modification on the road performances and thermal performance of asphalt binder and mixture. *Constr. Build. Mater.* **2019**, *220*, 64–75. [CrossRef]
12. Du, Y.; Dai, M.; Deng, H.; Deng, D.; Cheng, P.; Ma, C. Incorporating hollow glass microsphere to cool asphalt pavement: Preliminary evaluation of asphalt mastic. *Constr. Build. Mater.* **2020**, *244*, 118380.
13. Sun, X.; Qin, X.; Li, S.; Zou, C.; Wang, C.; Wang, X. Characterization of thermal insulating micro-surfacing modified by inorganic insulating material. *Constr. Build. Mater.* **2018**, *175*, 296–306. [CrossRef]
14. Du, Y.; Wang, S.; Zhang, J. Cooling asphalt pavement by a highly oriented heat conduction structure. *Energy Build.* **2015**, *102*, 187–196.
15. Zhu, Q.; Wang, W.; Wang, S.; Zhou, X.; Liao, G.; Wang, S.; Chen, J. Unilateral heat-transfer asphalt pavement for permafrost protection. *Cold Reg. Sci. Technol.* **2012**, *71*, 129–138. [CrossRef]

16. Wu, C.H.; Huang, C.H.; Li, Y.F.; Lee, W.H.; Cheng, T.W. Utilization of basic oxygen furnace slag in geopolymeric coating for passive radiation cooling application. *Sustainability* **2020**, *12*, 3967. [CrossRef]
17. Li, Q.; Ding, H.; Rahman, A.; He, D. Evaluation of Basic Oxygen Furnace (BOF) material into slag-based asphalt concrete to be used in railway substructure. *Constr. Build. Mater.* **2016**, *115*, 593–601. [CrossRef]
18. Chen, S.; Lin, D.; Luo, H.; Lin, Z. Application of reclaimed basic oxygen furnace slag asphalt pavement in road base aggregate. *Constr. Build. Mater.* **2017**, *157*, 647–653. [CrossRef]
19. Huang, L.S.; Lin, D.F.; Luo, H.L. Effect of field compaction mode on asphalt mixture concrete with basic oxygen furnace slag. *Constr. Build. Mater.* **2012**, *34*, 16–27. [CrossRef]
20. Xie, J.; Wu, S.; Lin, J.; Cai, J.; Chen, Z.; Wei, W. Recycling of basic oxygen furnace slag in asphalt mixture: Material characterization & moisture damage investigation. *Constr. Build. Mater.* **2012**, *36*, 467–474.
21. Brencis, G.; Smirnovs, J.; Zaumanis, M.; Haritonovs, V. Performance characterization of bituminous mixtures with dolomite sand waste and BOF steel slag. *J. Test. Eval.* **2012**, *40*, 1–6.
22. Lin, D.; Chou, L.; Wang, Y.; Luo, H. Performance evaluation of asphalt concrete test road partially paved with industrial waste—Basic oxygen furnace slag. *Constr. Build. Mater.* **2015**, *78*, 315–323. [CrossRef]
23. Shen, D.H.; Wu, C.M.; Du, J.C. Laboratory investigation of basic oxygen furnace slag for substitution of aggregate in porous asphalt mixture. *Constr. Build. Mater.* **2009**, *23*, 453–461. [CrossRef]
24. Amelian, S.; Manian, M.; Abtahi, S.M.; Goli, A. Moisture sensitivity and mechanical performance assessment of warm mix asphalt containing by-product steel slag. *J. Clean. Prod.* **2018**, *176*, 329–337. [CrossRef]
25. ASTM D3515. *Standard Specification for Hot-Mixed, Hot-Laid Bituminous Paving Mixtures*; ASTM: West Conshohocken, PA, USA, 1996.
26. ASTM D6926. *Standard Practice for Preparation of Asphalt Mixture Specimens Using Marshall Apparatus*; ASTM: West Conshohocken, PA, USA, 2020.
27. ASTM D6927. *Standard Test Method for Marshall Stability and Flow of Asphalt Mixtures*; ASTM: West Conshohocken, PA, USA, 2015.
28. ASTM D6931. *Standard Test Method for Indirect Tensile (IDT) Strength of Asphalt Mixtures*; ASTM: West Conshohocken, PA, USA, 2017.
29. ASTM E303-93. *Standard Test Method for Measuring Surface Frictional Properties Using the British Pendulum Tester*; ASTM: West Conshohocken, PA, USA, 2018.
30. Department of Traffic and Road in Queensland: Traffic and Road Management Manual, TRUM Vol 3, Part 2: Pavement Marking Usage. Available online: https://www.tmr.qld.gov.au/business-industry/Technical-standards-publications/Traffic-and-Road-Use-Management-manual/Volume-3 (accessed on 9 November 2020).

sustainability

Article

Influence of Carbonization Process on the Mechanical Properties of Nano-MgO Modified Cement Soil

Wei Wang [1], Hang Zhou [1], Jian Li [1], Feifei Tao [2], Cuihong Li [1], Biao Qian [3] and Ping Jiang [1,*]

[1] School of Civil Engineering, Shaoxing University, Shaoxing 312000, China; wellswang@usx.edu.cn (W.W.); zhouhang1119@126.com (H.Z.); lijianusx@163.com (J.L.); lich@usx.edu.cn (C.L.)
[2] School of Chemistry and Chemical Engineering, Shaoxing University, Shaoxing 312000, China; feifeitao@usx.edu.cn
[3] Tongchuang Engineering Design Co. Ltd., Shaoxing 312000, China; tcgcsj@tcgc.ltd
* Correspondence: jiangping@usx.edu.cn

Abstract: In order to explore the modification effect of carbonization time on nano-MgO-modified cement soil, unconfined compressive strength tests of nano-MgO-modified cement soil with carbonization times of 0 h, 6 h, 1 d, 2 d and 4 d were carried out. A method for normalizing the stress–strain curve was proposed, and the influence of nano-MgO content and carbonization time was investigated from the three aspects of compressive strength, peak strain and energy dissipation. The test results show the following: (1) The compressive strength of the modified cement soil can be significantly improved by adding 1.0% nano-MgO and after 1 d carbonization. (2) Under the same nano-MgO content, the peak strain of the modified cement soil after 2 d carbonization reaches the maximum, which can significantly increase its ductility. However, the nano-MgO content has little influence on the peak strain of the modified cement soil. (3) Under the same nano-MgO content, the energy dissipation rate of the modified cement soil after 1 d carbonization reaches the maximum, which can better resist the damage of external load.

Keywords: cement soil; nano MgO; carbonization process; compressive strength; energy dissipation

Citation: Wang, W.; Zhou, H.; Li, J.; Tao, F.; Li, C.; Qian, B.; Jiang, P. Influence of Carbonization Process on the Mechanical Properties of Nano-MgO Modified Cement Soil. *Sustainability* **2021**, *13*, 3558. https://doi.org/10.3390/su13063558

Academic Editor: Yeou-Fong Li

Received: 10 February 2021
Accepted: 21 March 2021
Published: 23 March 2021

1. Introduction

Nowadays, with the rapid development of infrastructure construction, there are fewer land resources available. In the process of building construction, soft soil foundations with lower strength are often encountered. Cement and lime [1–4] are often used to improve the bearing capacity of soft soil foundations to make them meet the needs of different engineering fields, such as highways, slopes and residential construction [5,6]. However, soft soil has many undesirable engineering properties, such as high compressibility and high porosity and obvious thixotropy and creep properties [7], which require a large amount of cement to solidify the soft soil foundation. Moreover, CO_2 produced by cement production accounts for 5–7% of the total amount of human-generated CO_2 [8]. Therefore, in order to reduce the dosage of cement and CO_2 emissions, many scholars have found suitable materials and methods to further improve cement soil on the basis of adding new engineering materials as a curing agent [9–12].

Nanomaterials are widely used in the field of construction engineering due to their characteristics of small particle size, large surface area and good stability. Commonly used nanomaterials are nano-SiO_2 [13–15], nano-clay [16–18] and nano-MgO. Among them, nano-MgO is mainly made by sintering magnesite at 700 °C, while the temperature of cement in the production process is above 1450 °C [19]. Therefore, using nano-MgO to replace part of cement can also reduce energy consumption [20]. Gao et al. [21,22] explored through tests that 6% nano-MgO content could significantly increase the compressive strength and stability of clay, and better improve the shear strength and cohesion of the modified cement soil. However, it had little influence on the internal friction angle.

Moradpour et al. [23] found that the compressive strength and flexural strength of cement-based materials was significantly improved with 1.0% nano-MgO content. In addition, nano-MgO had expansibility, and the microstructure of nano-MgO-modified cement-based materials was denser than the ordinary cement-based materials. Wang et al. [24] found that under the erosion of sulfuric acid solution, the shear strength of modified cement soil reached the maximum with 0.5% nano-MgO content. Yao et al. [25,26] investigated through unconfined tests that the addition of nano-MgO could better improve the strength of cement soil. Moreover, the cohesion of modified cement soil reached the maximum, with the 1.0% nano MgO content, and the internal microstructure was more compact. Yuan et al. [27] and Hou et al. [28] found that the addition of nano-MgO would form a large amount of fluffy and acicular aluminate hydration products inside the cement paste, which could effectively fill the larger pores inside the cement paste and improve the stability of the structure.

As CO_2 will lead to the corrosion of steel bars, research mainly focuses on the prevention and control measures of the carbonation of concrete [29–31]. However, the corrosion of steel bars is not considered in solidified soft soil, so CO_2 can be injected into solidified soft soil to improve its engineering performance. Moreover, CO_2 could react with hydration products to consume CO_2 and reduce greenhouse gas emission. Yi et al. [32] found that after carbonization, the unconfined compressive strength of solidified soil with 5% MgO content was about 200% of that of ordinary cement solidified soil. However, with 10% MgO content, the increase in strength decreased. Cai et al. [33] found that with the increase in carbonization time, the number of cracks on the surface of MgO solidified sand gradually increased. Vandeperre et al. [34] found that MgO generated rankinite after carbonization, which had a larger volume expansion rate and could fill the pores of the mixture, so as to improve the strength and toughness of the mixture. Mo et al. [35] found that under the action of carbonization, the pore size and porosity of MgO-modified cement paste decreased, while the apparent density and microscopic density increased.

In sum, the research on the mechanical properties of nano-MgO-modified cement soil has been gradually carried out. However, there are few studies on the mechanical properties of the modified cement soil by carbonization, which requires further exploration. In this paper, on the basis of cement soil with 20% mass fraction of cement, the unconfined compressive strength tests were carried out on the modified cement soil with different nano-MgO contents and carbonized times. From the angle of strength and ductility, the influence of carbonization time on the mechanical properties of the nano-MgO-modified cement soil was investigated.

2. Test Materials and Preparation

2.1. Materials

The materials used throughout the experimental tests include subgrade soil, water, cement and nano-MgO. The subgrade soil was collected from a coastal area in Shaoxing city of Zhejiang province, China. The main physical and mechanical indexes are shown in Table 1. Guyue brand P.C 42.5 composite Portland cement produced by Shaoxing Keqiao Third Cement Co., Ltd. (Shaoxing City, China) was used as shown in Figure 1a, and the main physical and mechanical parameters are shown in Table 2. The nano-MgO was a 50 nm spherical MgO produced by Shanghai Macklin Biochemical Technology Co., Ltd (Shanghai City, China). The appearance was white to light yellow powder, as shown in Figure 1b, and its physical and mechanical properties are shown in Table 3.

Table 1. The physical indexes of subgrade soil in coastal area of Shaoxing.

Density (ρ/g·cm^{-3})	Void Ratio e	Saturability (%)	Natural Moisture Content w (%)	Liquid Limit W_L (%)	Plastic Limit W_P (%)	Liquidity Index I_L (%)	Plasticity Index I_P (%)
1.65	1.64	98	60.00	46.20	26.40	1.70	19.80

(a) (b)

Figure 1. Physical picture of cement and nano MgO. (**a**) Composite Portland cement; (**b**) nano-MgO.

Table 2. The physical and mechanical parameters of P.C 42.5 cement.

Fineness(%)		Stability	Setting Time (min)		Flexural Strength (MPa)		Compressive Strength (MPa)	
0.08 mm ≤10	0.045 mm ≤30	Qualified	Initial ≥45	Final ≤600	3 d ≥2.5	28 d ≥5.5	3 d ≥10.0	28 d ≥42.5

Table 3. The physical and mechanical parameters of nano MgO.

Mean Grain Size (nm)	Purity (%)	Melting Point (°C)	Boiling Point (°C)	Density (20 °C) ($g{\cdot}cm^{-3}$)	Specific Surface Area ($m^2{\cdot}g^{-1}$)	Refractive Index (%)
50	99.9	2852	3600	3.580	30–50	1.736

2.2. Test Program

The samples were divided into 25 groups according to different nano-MgO contents and carbonization times, and the mix proportion of each group of materials is shown in Table 4. MCS represents nano-MgO-modified cement soil. Unconfined compressive strength tests were carried out on 25 groups of MCS. In order to reduce the random error and the discreteness of the test data in the test process, five repeated tests were performed on each group of samples.

Table 4. The sample mix proportion and carbonation time design.

Group	Cement Content (%)	Moisture Content (%)	Curing Time (d)	Nano-MgO Content (%)	Carbonization Time
MCS	20	80	7	0, 0.5, 1.0, 1.5, 2.0	0 h, 6 h, 1 d, 2 d, 4 d

2.3. Sample Preparation

(1) Soil sample preparation. The subgrade soil was soaked in water for 7 days until it was softened and turned into silt. Next, large particles, stones or other impurities were removed from the subgrade soil through a sieve with a diameter of 2 mm.

(2) Determination of the moisture content. The subgrade soil was stirred evenly and stood for 2–3 weeks. Then, a small amount of subgrade soil was taken out to measure its moisture content.

(3) Sample mixing. Appropriate amounts of subgrade soil, cement and nano-MgO were weighed according to the experimental mix proportion, and placed into a mixer and stirred for 5 min.

(4) Unconfined sample preparation. The prepared mixture was poured into a cylindrical mold with a diameter of 39.1 mm and a height of 80 mm three times. Each time

the mixture was poured, it was vibrated. The sample was left to stand for 2 h after completion, and the upper and lower surfaces of the samples were flattened with a scraper, wrapped with filter paper and placed into water for curing.

(5) Sample curing. The moisture content of the sample was high, so it was difficult to form; therefore, it needed to soak in water for 3 days before demolding. Then, it was placed into a standard curing box or carbonization curing box to continue curing. The curing methods and conditions are shown in Figure 2 and Table 5, and the soaking and carbonization curing are shown in Figure 3a,b, respectively.

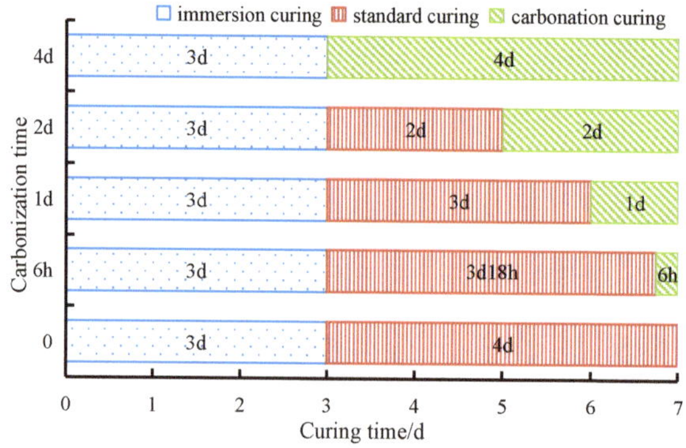

Figure 2. The relationship between carbonation time and curing age.

Table 5. The maintenance conditions.

Maintenance Method	Maintenance Conditions
immersion curing	water temperature (20 °C ± 2 °C)
standard curing	temperature (20 °C ± 2 °C); humidity (95% ± 2%)
carbonization curing	temperature (20 °C ± 2 °C); humidity (95 ± 2%); CO_2 concentration (20% ± 3%)

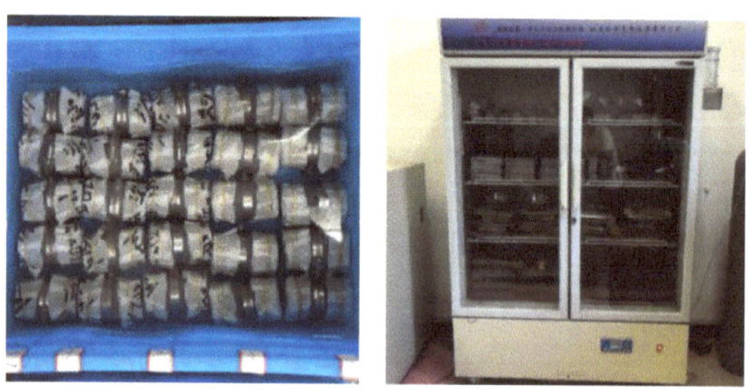

(**a**) (**b**)

Figure 3. Maintenance method. (**a**) Immersion curing; (**b**) carbonization curing.

2.4. Mechanical Test

TKA-WXY-1F automatic unconfined pressure gauge produced by Nanjing TKA Technology Co., Ltd. was used in the unconfined compressive strength test. The loading rate was set to 1 mm/min during the test. According to the GB/T 50123-2019 Geotechnical Test Standard, when the axial force reaches its peak value, the test can be stopped by performing another 3–5% axial strain value. In this paper, the test was stopped when the axial strain of uncarburized MCS reached 10%. The test was stopped when the axial strain of carburized MCS reached 12%.

3. Unconfined Data Analysis

3.1. Stress–strain Curve Analysis

The stress–strain curves of MCS with different nano-MgO contents and carbonization times are shown in Figures 4–8. It can be seen from the figures that the stress–strain curves of MCS are all softening curves. At the initial stage of loading, the stress of MCS increases linearly with the increase in strain. As the strain continued to increase, the upward trend of stress gradually slowed down, and finally, the stress reaches the peak value. At this point, the strain corresponding to the peak stress is the peak strain of the sample. When the peak strain occurs, the sample will be destroyed and inclined cracks will appear, and the force on the sensor will drop sharply, so the stress attenuation rate will increase. With the increase in strain, the sample will be compressed, and the friction force will be generated at the inclined cracks, so the stress attenuation rate will decrease. With the continuous increase in strain, the sample will be further compressed, and the force on the sensor will not change.

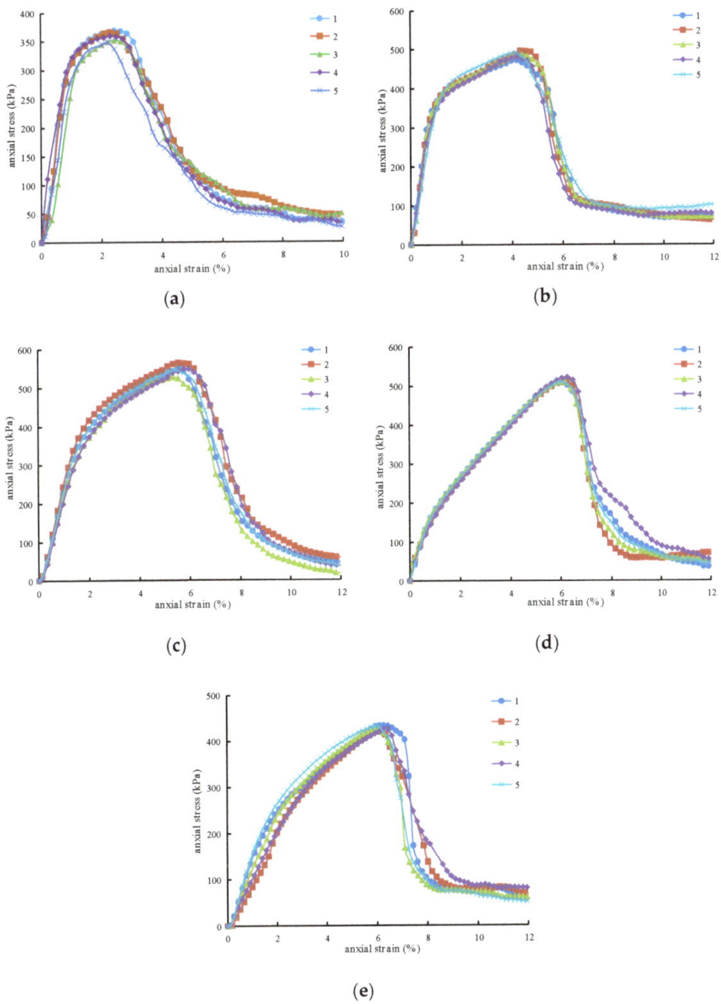

Figure 4. The stress–strain curve of MCS with 0% nano-MgO content. (**a**) Carbonization 0 d; (**b**) carbonization 6 h; (**c**) carbonization 1 d; (**d**) carbonization 2 d; (**e**) carbonization 4 d.

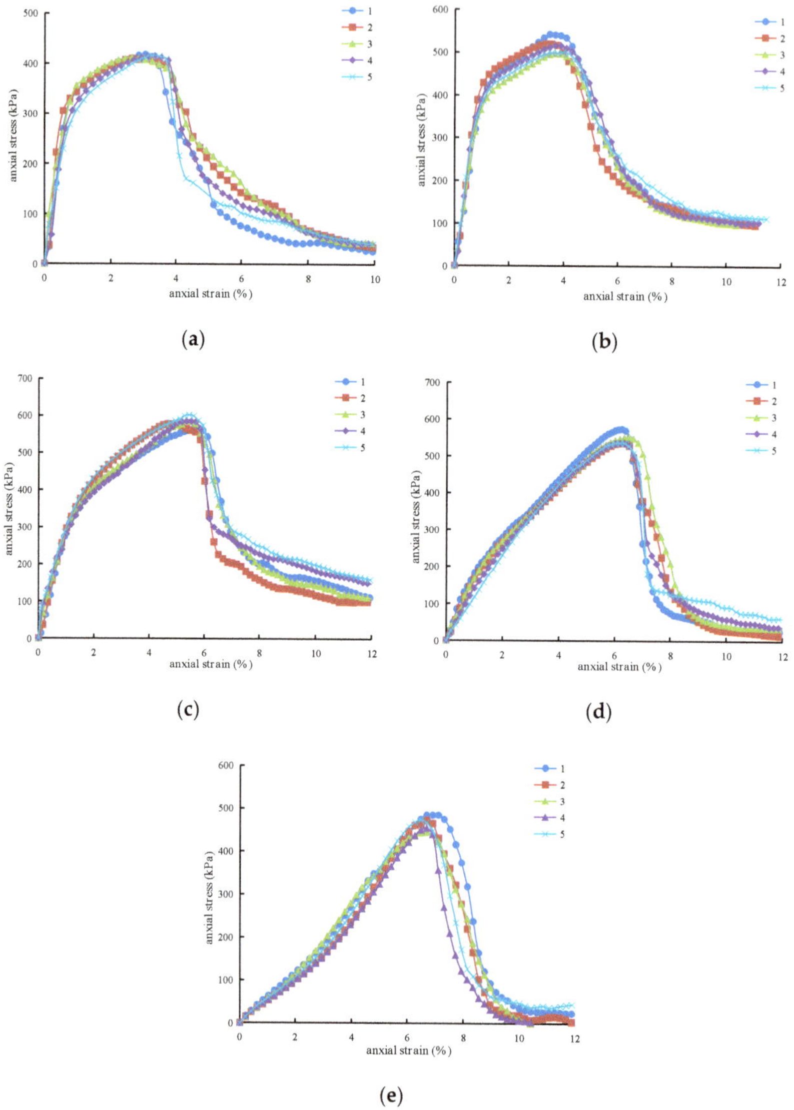

Figure 5. The stress–strain curve of MCS with 0.5% nano-MgO content. (**a**) Carbonization 0 d; (**b**) carbonization 6 h; (**c**) carbonization 1 d; (**d**) carbonization 2 d; (**e**) carbonization 4 d.

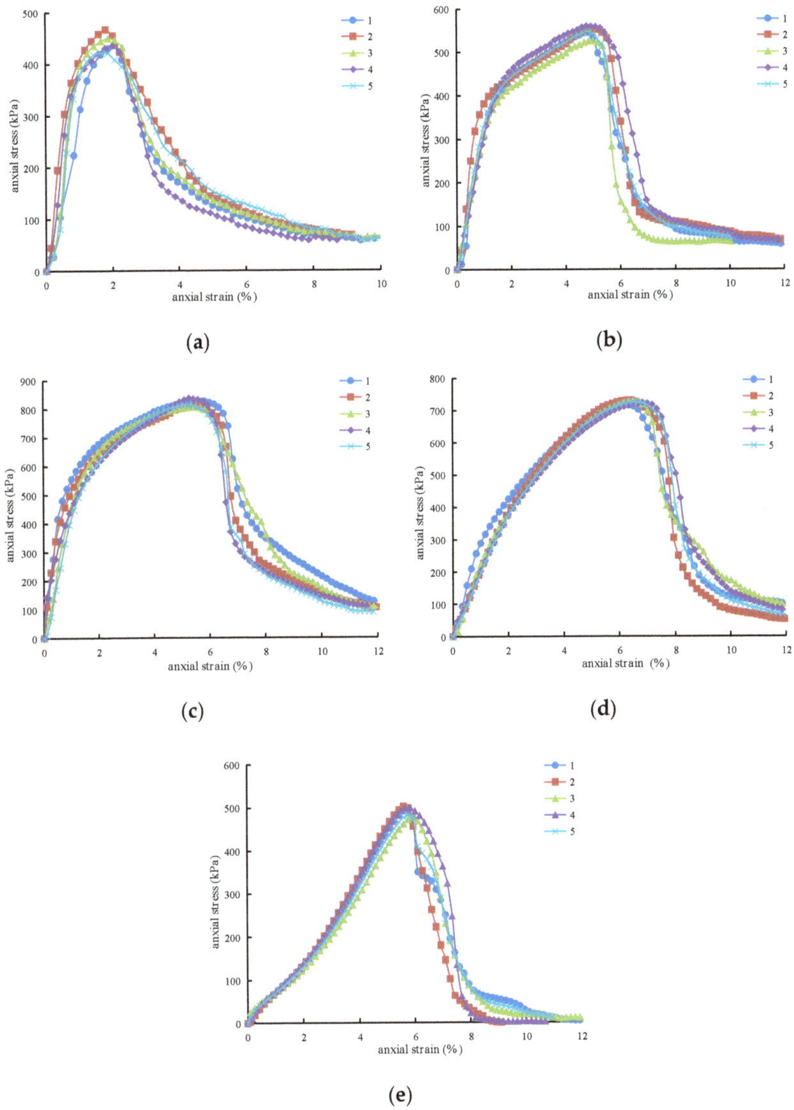

Figure 6. The stress–strain curve of MCS with 1.0% nano-MgO content. (**a**) Carbonization 0 d; (**b**) carbonization 6 h; (**c**) carbonization 1 d; (**d**) carbonization 2 d; (**e**) carbonization 4 d.

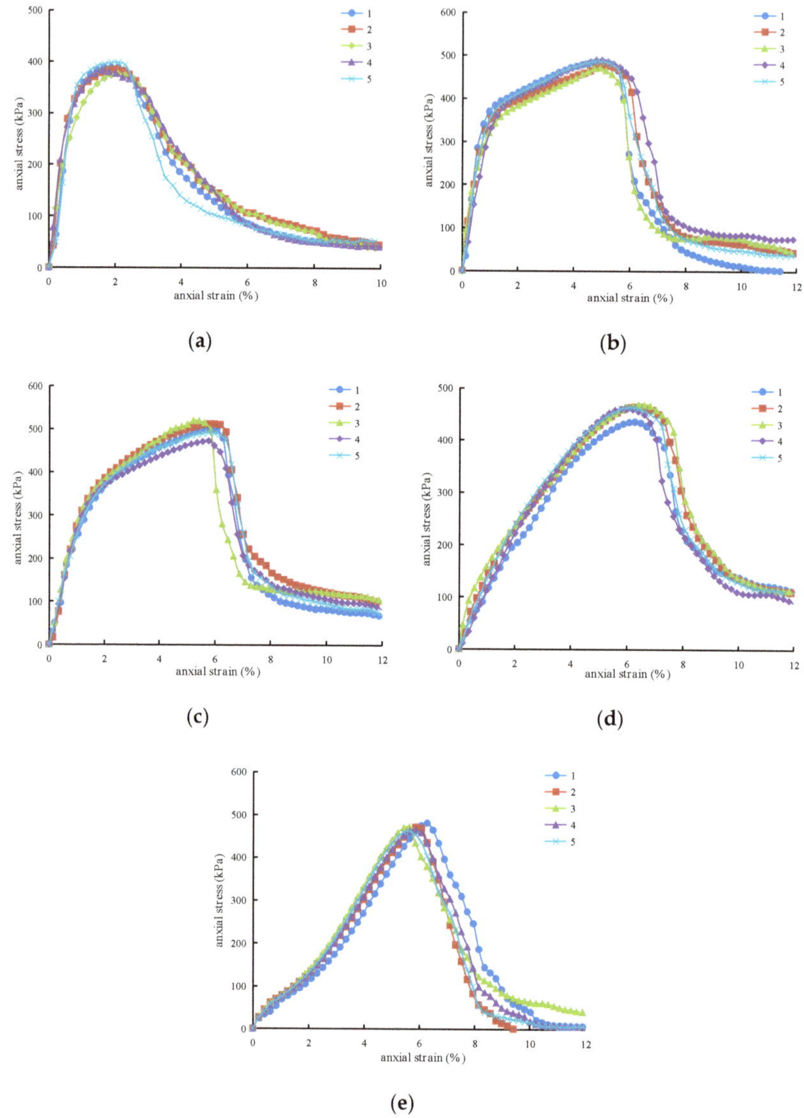

Figure 7. The stress–strain curve of MCS with 1.5% nano-MgO content. (**a**) Carbonization 0 d; (**b**) carbonization 6 h; (**c**) carbonization 1 d; (**d**) carbonization 2 d; (**e**) carbonization 4 d.

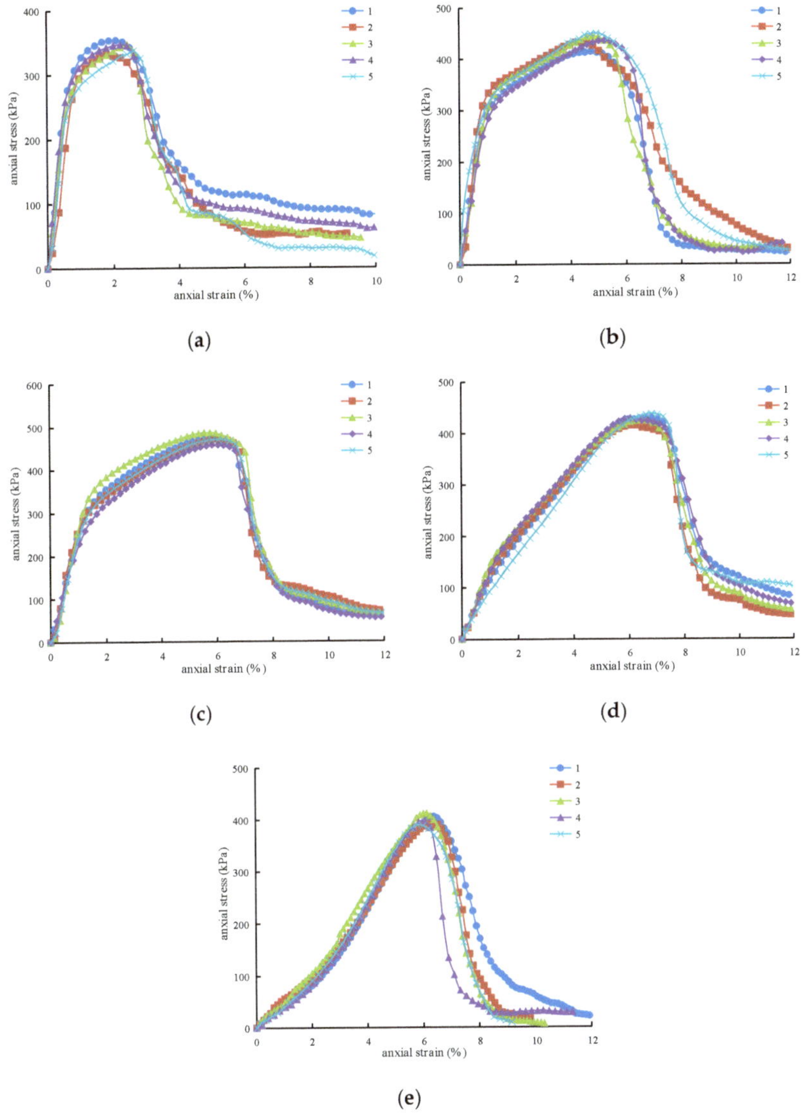

Figure 8. The stress–strain curve of MCS with 2.0% nano-MgO content. (**a**) Carbonization 0 d; (**b**) carbonization 6 h; (**c**) carbonization 1 d; (**d**) carbonization 2 d; (**e**) carbonization 4 d.

3.2. *Curve Normalization*

During the test, five repeated tests were conducted for all the tests designed in Table 4 to reduce the test error. The results of five repeated samples are different, so obtaining a reasonable standard value from these five results is particularly critical. In this paper, the deviation of five peak stresses were taken as the research object, and the weight of each peak stress was calculated. Then, the weight of each peak stress was taken as the weight of this curve, and the five curves were further normalized to one curve. The normalization method is as follows:

1. Determination of the average value of peak stress? $\overline{\sigma}_i$.

$$\overline{\sigma} = \frac{1}{N} \sum_{i \in [1,N]} \sigma_i \tag{1}$$

where σ_i is the peak stress of each stress–strain curve of the sample, and N is the number of repeated tests in this group of samples. In this study, there were five repeated tests in each group, so $N = 5$, and i is the serial number of the data in the sample.

2. Determination of the deviation $\Delta\sigma_i$ of each peak stress.

$$\Delta\sigma_i = |\sigma_i - \overline{\sigma}| \tag{2}$$

3. Determination of the variance S of the peak stress.

$$s = \frac{1}{N} \sum_{i \in [1,N]} \Delta\sigma_i^2 \tag{3}$$

4. Determination of the initial weight p_i of each peak stress. In order to ensure that the initial weight is between [0,1], the maximum value $\Delta\sigma_{imax}$ of peak stress deviation is introduced to calculate the initial weight of each peak stress.

$$p_i = \frac{\Delta\sigma_i}{\Delta\sigma_i \max + S} \tag{4}$$

5. Determination of the weight mapping function $f(x)$. In the above equations, the greater the deviation of the peak value is, the greater the weight of the peak stress is. Therefore, it is necessary to assign the weight of peak stress, so that the larger the deviation of the peak stress is, the smaller the weight after assignment is. The mapping function is shown in Equation (5). As the function decreases monotonously, the greater the $\Delta\sigma_i$ is, the smaller the weight after conversion is.

$$f(x) = \frac{2}{\pi} \times \arccos(x) \tag{5}$$

6. Determination of the converted weight m_i of each peak stress.

$$m_i = f(p_i) \tag{6}$$

7. Determination of the weighting factor n_i of each peak stress.

$$n_i = \frac{m_i}{\sum_{i \in [1,N]} m_i} \tag{7}$$

8. Determination of the standard value σ of peak stress.

$$\sigma = \sum_{i \in [1,N]} \sigma_i \times n_i \tag{8}$$

Five curves of MCS with 1.0% nano-MgO after 6 h carbonization were taken as examples and weighted by the above methods. Five test curves and the weighted curves are shown in Figure 9. It can be seen from the figure that the normalized curve has a good correlation with the original five test curves.

Five stress–strain curves of MCS with different nano MgO contents and carbonization times were normalized by the above method. The normalized curve is shown in Figure 10.

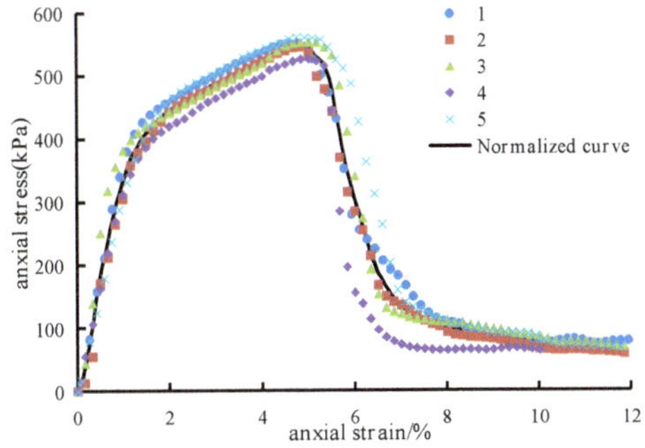

Figure 9. Weighted curve of MCS with 1.0% nano-MgO content after 6 h carbonization.

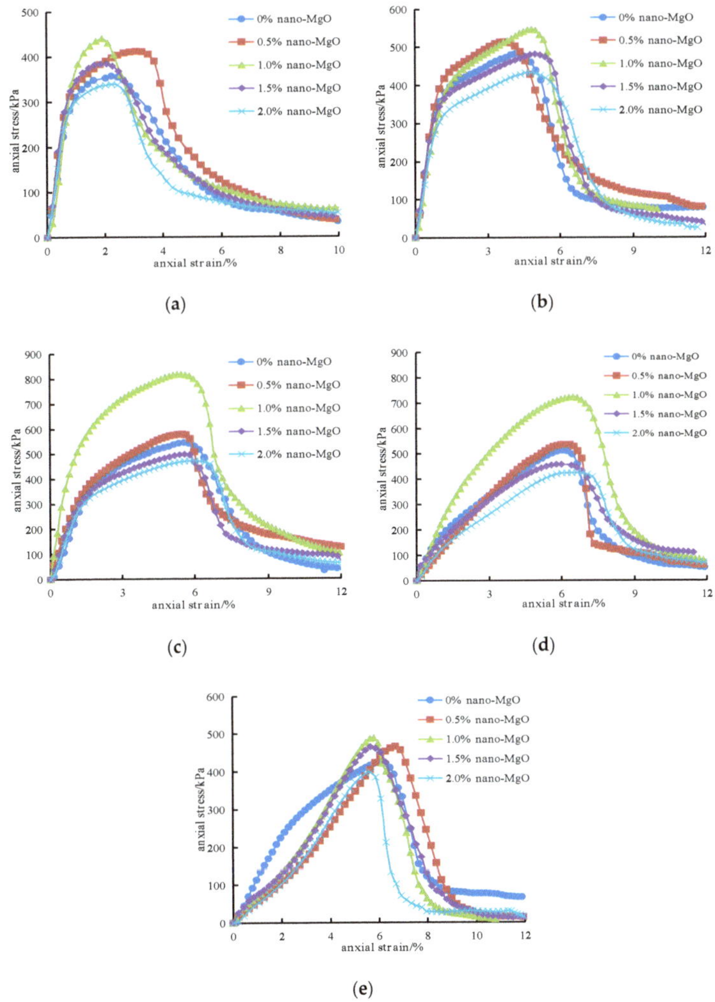

Figure 10. The Normalized stress–strain curves of MCS. (**a**) Carbonization 0 d; (**b**) carbonization 6 h; (**c**) carbonization 1 d; (**d**) carbonization 2 d; (**e**) carbonization 4 d.

4. Discussion

4.1. The Influence of Nano-MgO Content

The maximum value of the normalization curve of each group of samples is taken as the peak stress of each group of samples. The compressive strength of MCS with different nano-MgO contents is shown in Figure 11. It can be seen from the figure that under the same carbonization time, the compressive strength of MCS increases first and then decreases with the increase in nano-MgO content. When the nano-MgO content is 1.0%, the compressive strength of MCS reaches the maximum. When the carbonization time is 0 h, 6 h, 1 d, 2 d and 4 d, the compressive strength is 440.69 Pa, 547.33 kPa, 818.25 kPa, 724.10 kPa and 489.18 kPa, respectively, which is 23.3%, 13.3%, 50.2%, 41.5% and 15.1% higher than that of cement soil without nano-MgO.

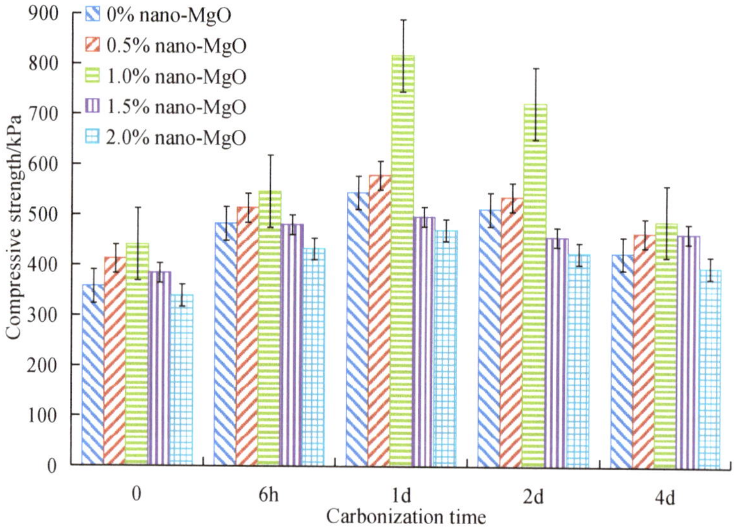

Figure 11. The influence of different nano-MgO contents on the compressive strength of MCS.

This is mainly because when the nano-MgO content is between 0% and 1.0%, cement and nano-MgO generate carbonate precipitation after hydration and carbonization, and the volume increases and gradually fills the pores inside MCS. As a result, the internal structure of the soil is denser. However, with the increase in nano-MgO content, the reaction generates excessive carbonate precipitation, and the volume expansion exceeds the volume of the pores inside MCS [25,36], which gradually compresses the surrounding soil, resulting in the failure of cementation between soil particles. Microcracks appear inside MCS, and the compressive strength decreases.

4.2. The Influence of Carbonization Time

The influence of different carbonization times on the compressive strength of MCS is shown in Figure 12. It can be seen from the figure that the compressive strength of MCS with the same nano-MgO content increases first and then decreases with the increase in carbonization time, so there is an optimal carbonization time. Under the five nano-MgO contents, the optimal carbonization time is 1 d. At this time, the compressive strength of MCS is 544.92 kPa, 579.36 kPa, 818.25 kPa, 497.55 kPa and 470.92 kPa, respectively, which is 52.5%, 40.5%, 85.7%, 29.5% and 38.8% higher than that of non-carbonized MCS. When the carbonization time is 4 d, the compressive strength of MCS under five nano-MgO contents is 425.01 kPa, 465.68 kPa, 489.18 kPa, 461.74 kPa and 397.26 kPa, respectively, which is still 18.9%, 12.9%, 11.0%, 20.2% and 17.08% higher than that of non-carbonized MCS.

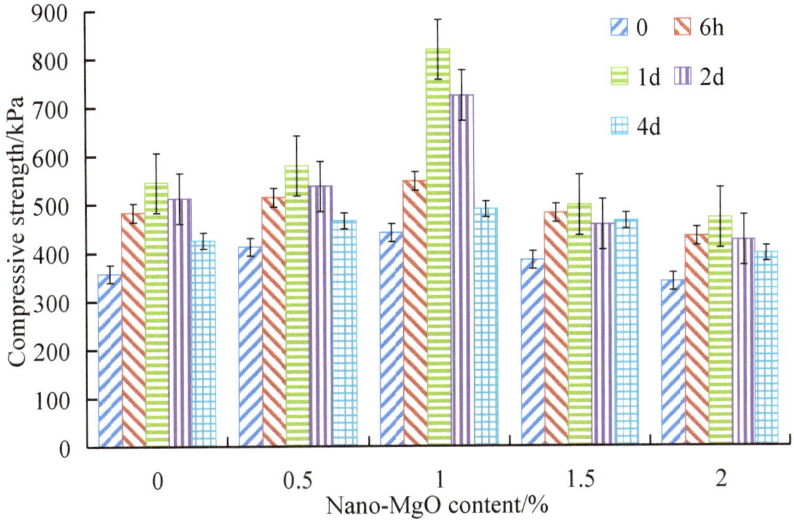

Figure 12. The influence of carbonization time on the compressive strength of MCS.

The mechanism of action is similar to the influence of nano-MgO content. When the carbonization time is short, the carbonate precipitation formed by carbonization can fill the internal pores. However, as the carbonization time increases, excessive carbonate precipitation causes cracks inside MCS, leading to a decrease in the compressive strength.

5. Ductility and Energy Dissipation Analysis

5.1. Curve Simplification

According to the change trend of the normalized stress–strain curve, the stress–strain curve of MCS can be divided into four stages, as shown in Figure 13:

(1) Elastic stage (OA section). At this stage, the pores inside the sample are compressed, the volume is reduced, and there are no cracks on the surface (Figure 14a). The contact between the soil particles is closer, and the structure becomes stronger. The stress increases linearly with the increase in strain and reaches the stress corresponding to strengthening point A.

(2) Strengthening stage (AB section). As the axial load increases, the sample undergoes plastic deformation, and small cracks appear on the surface (Figure 14b). At this stage, the stress growth trend of the sample slows down and gradually reaches the stress corresponding to the peak point B.

(3) Falling stage (BC section). As the axial load continues to increase, the sample shows uneven plastic deformation, and the cracks on the surface gradually spread expand to the surroundings (Figure 14c). At this stage, as the strain increases, the stress begins to decrease, and the downward trend gradually increases until it drops to the stress at point A, at which point, the stress–strain corresponds to the drop point C.

(4) Failure stage (CD section). As the axial load continues to increase, the cracks on the surface gradually expand to the surroundings and are interconnected to one another to form a penetrating crack, accompanied by soil shedding (Figure 14d). The stress decreases rapidly with the increase in strain, but the downward trend gradually slows down and finally approaches a fixed value. At this time, the point corresponding to the stress and strain is the failure point D.

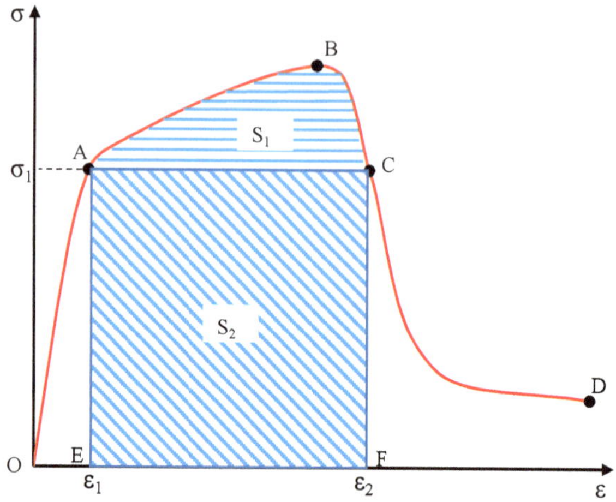

Figure 13. The change trend of stress–strain curve.

(**a**) (**b**) (**c**) (**d**)

Figure 14. Failure mode. (**a**) OA section; (**b**) AB section; (**c**) BC section; (**d**) CD section.

5.2. Ductility Analysis

The strain at point B of the normalized curve of each group of samples was taken as the peak strain of the group. The influence of different carbonization times on the peak strain of MCS is shown in Figure 15. It can be seen from the figure that when the nano-MgO content is 0 and 0.5%, the peak strain of MCS increases with the increase in carbonization time. However, when the carbonization time increases from 0 to 2 d, the peak strain of MCS increases significantly, and the difference between the peak strain of MCS and the non-carbonized MCS is 3.8% and 3.0%, respectively; when the carbonization time increases from 2 d to 4 d, the peak strain increases by 0.1% and 0.6%, respectively. Therefore, the effect of 2 d carbonization on the peak strain of MCS is more obvious, which can significantly increase the ductility of MCS.

When the nano-MgO content is 1.0%, 1.5% and 2.0%, the peak strain of MCS first increases and then decreases with the increase in carbonization time. When the carbonization time is 2 d, the peak strain reaches the maximum, which is 6.5%, 6.2% and 6.7%, respectively, and the difference between the peak strain of MCS and non-carbonized MCS is 4.6%, 4.2% and 4.4%, respectively.

When the carbonization time is 0 h, 6 h, 1 d, 2 d and 4 d, the difference between the maximum and minimum peak strains of MCS with different nano-MgO contents is 1.2%,

1.3%, 0.7%, 0.6% and 1.1%, respectively. It can be found that the influence of different nano-MgO contents on the peak strain is far less significant than that of carbonization time.

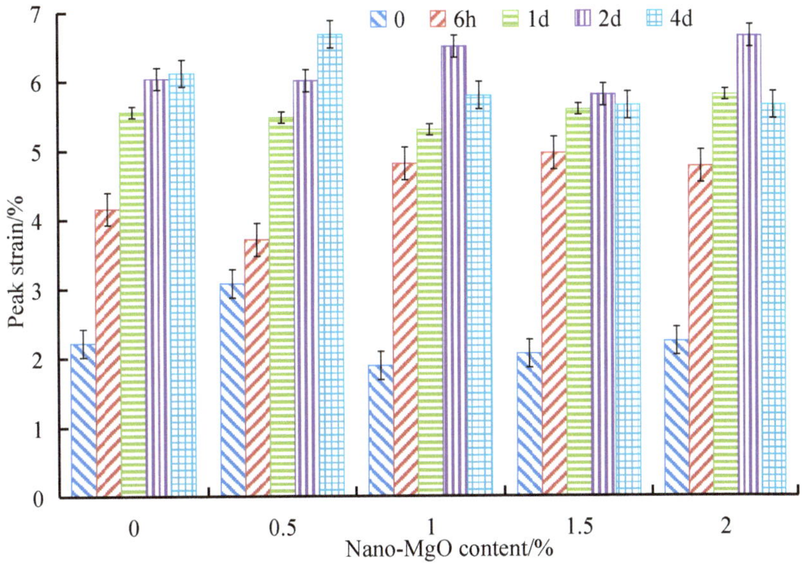

Figure 15. The influence of carbonation time on the peak strain of MCS.

Therefore, under the same nano-MgO content, 2 d carbonization has the best effect on improving the peak strain of MCS, which can significantly increase the ductility of MCS. Under the same carbonization time, the nano-MgO content has little effect on the peak strain of MCS.

5.3. Energy Dissipation Analysis

During the loading process, energy dissipation leads to the process of the formation of internal cracks. In the elastic stage, there are almost no cracks inside MCS, so the energy storage and release are reversible. In the strengthening stage, damage and cracks occur inside MCS, leading to energy dissipation. The greater the energy dissipation is, the stronger the ability of MCS to resist damage is. Therefore, energy dissipation can be used to measure the ability of MCS to resist damage.

As shown in Figure 13, the area enclosed by ABCFE is the total energy dissipation density S of the sample; the area enclosed by ABC is the enhanced energy dissipation density S_1 of the sample; the area enclosed by ACFE is the basic energy dissipation density S_2 of the sample. The calculation formula is shown in Equation (9), and the unit is kJ/m^3.

$$\begin{cases} S & = \int_{\varepsilon_1}^{\varepsilon_2} f(\varepsilon)d\varepsilon \\ S_2 & = \sigma_1 \times (\varepsilon_2 - \varepsilon_1) \\ S_1 & = S - S_2 \end{cases} \tag{9}$$

σ_1 is the stress at point A on the $f(x)$ curve; ε_1 and ε_2 are the strains at point A and point C on the $f(x)$ curve, respectively. As shown in Figure 10, the normalized curve of each group of samples can be obtained by using the above curve normalization method, so as to better determine the specific location of points A, B, C and D. In order to better measure the degree of energy dissipation of MCS, the ratio of S_1 and S is taken as the energy dissipation rate K of MCS to reflect the ability to resist external damage. The larger the K is, the stronger the ability of MCS to resist external damage is. The influence of different carbonization times on the energy dissipation rate of MCS is shown in Figure 16.

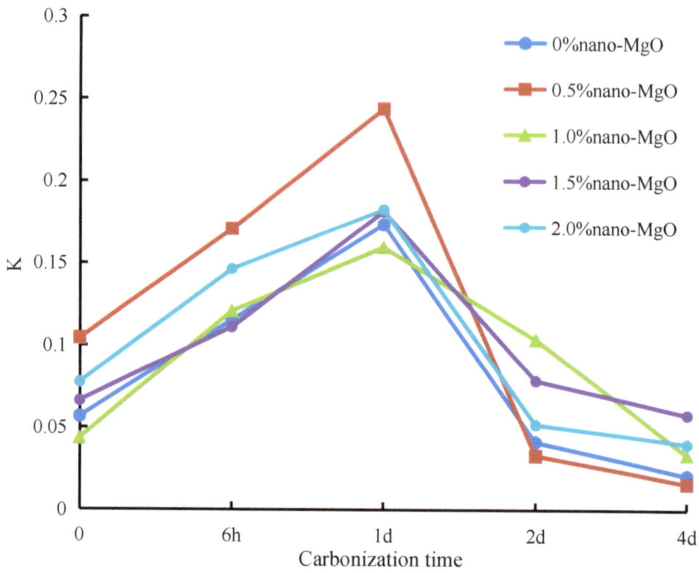

Figure 16. The influence of carbonization time on the energy dissipation rate of MCS.

It can be seen from Figure 16 that the K of MCS increases first and then decreases with the increase in carbonization time under the same content of nano-MgO. When the carbonization time is 1 d, the K of MCS with five nano-MgO content reaches the maximum, which is 0.17, 0.24, 0.16, 0.18 and 0.18, respectively. This is much higher than that of non-carbonized MCS. Therefore, under five nano-MgO contents, the MCS after 1 d carbonization can better resist external damage.

When the carbonization time is 0 h, 6 h and 1 d, the K of MCS with 0.5% nano-MgO content is the largest, which is 0.10, 0.17 and 0.24, respectively. When the carbonization time is 2 d, the K of MCS with 1.0% nano-MgO content reaches the maximum value, 0.10. When the carbonization time is 4 d, the K of MCS under five nano-MgO contents is 0.02, 0.02, 0.03, 0.05 and 0.04, respectively, and the range of change is small. Therefore, when the carbonization time is 4 d, different nano-MgO contents have no significant influence on the K of MCS.

In summary, MCS with 0.5% nano-MgO content has the largest energy dissipation rate after 1 d carbonization, which is much greater than the other MCS with other nano-MgO contents and after other carbonization times. This indicates that MCS with 0.5% nano-MgO content can better resist external damage after 1 d carbonization.

6. Conclusions and Discussion

6.1. Conclusions

Through the unconfined compressive strength tests, the influence of different nano-MgO contents and carbonization times on MCS was explored from the three aspects of compressive strength, peak strain and energy dissipation. The following conclusions can be drawn:

(1) A method for normalizing the stress–strain curve was proposed, and the normalized curve had a good correlation with the five stress–strain curves obtained through the tests.

(2) Under the same carbonization time, the compressive strength of MCS first increases and then decreases with the increase in nano-MgO content. When the nano-MgO content is 1.0%, the compressive strength of MCS reaches the maximum, which is 23.4%, 13.3%, 50.2%, 41.4% and 15.0% higher than that of cement soil without nano-MgO. Under the same nano-MgO content, the compressive strength of MCS first increases and then decreases with the increase in carbonization time under five nano-MgO contents. Therefore, when the nano-MgO content is 1.0%, the compressive strength of MCS after 1 d carbonization reaches the maximum.
(3) When the nano-MgO content is 0% and 0.5%, the peak strain of MCS increases with the increase in carbonization time. When the carbonization time increases from 2 d to 4 d, the increase in peak strain is smaller, so the modification effect of 2 d carbonization is optimal. When the nano-MgO content is 1.0%, 1.5% and 2.0%, the peak strain of MCS first increases and then decreases with the increase in carbonization time. When the carbonization time is 2 d, the peak strain reaches the maximum. Therefore, under five nano-MgO contents, 2 d carbonization can significantly increase the peak strain of MCS and its ductility. However, the nano-MgO content has little influence on the peak strain of MCS.
(4) Under the same nano-MgO content, the energy dissipation rate of MCS first increases and then decreases with the increase in carbonization time. When the carbonization time is 1 d, the energy dissipation rate of MCS reaches the maximum under five nano-MgO contents, which can better resist the damage of external load. However, when the carbonization time is 4 d, the energy dissipation rate of MCS is the smallest, and different nano-MgO contents have almost no effect on the energy dissipation rate.

6.2. Discussion

(1) In this paper, the CO_2 concentration during the carbonization process is 20% ± 3%, and the influence of CO_2 concentration on MCS is not considered.
(2) During the curing process, the samples were carbonized in the later period of the curing, and the effect of carbonization at the early and middle stages of the curing on the test results was not explored in depth.

Author Contributions: Conceptualization, B.Q. and P.J.; formal analysis, F.T. and C.L.; investigation, H.Z. and J.L.; writing—review and editing, W.W. All authors have read and agreed to the published version of the manuscript.

Funding: This research was funded by the National Natural Science Foundation of China (grant number 41772311), the Scientific Research Projects of Zhejiang Department of Housing and Urban and Rural Construction of China (grant number 2017K179, 2019K171) and the International Scientific and Technological Cooperation Projects of Shaoxing University (grant number 2019LGGH1007).

Institutional Review Board Statement: Not applicable.

Informed Consent Statement: Not applicable.

Data Availability Statement: The data presented in this study are available on request from the corresponding author.

Conflicts of Interest: The authors declare no conflict of interest.

References

1. Liu, Y.; He, L.Q.; Jiang, Y.J.; Sun, M.M.; Chen, E.J.; Lee, F.-H. Effect of in situ water content variation on the spatial variation of strength of deep cement-mixed clay. *Géotechnique* **2019**, *69*, 391–405. [CrossRef]
2. Federico, A.; Vitone, C.; Murianni, A. On the mechanical behaviour of dredged submarine clayey sediments stabilized with lime or cement. *Can. Geotech. J.* **2015**, *52*, 2030–2040. [CrossRef]
3. Wang, W.; Li, J.; Hu, J. Unconfined Mechanical Properties of Nanoclay Cement Compound Modified Calcareous Sand of the South China Sea. *Adv. Civ. Eng.* **2020**, *2020*, 1–16. [CrossRef]
4. Jha, A.K.; Sivapullaiah, P. Susceptibility of strength development by lime in gypsiferous soil—A micro mechanistic study. *Appl. Clay Sci.* **2015**, *115*, 39–50. [CrossRef]

5. Xu, B.; Low, B.K. Probabilistic Stability Analyses of Embankments Based on Finite-Element Method. *J. Geotech. Geoenviron. Eng.* **2006**, *132*, 1444–1454. [CrossRef]
6. Liu, Y.; Zhang, W.; Zhang, L.; Zhu, Z.; Hu, J.; Wei, H. Probabilistic stability analyses of undrained slopes by 3D random fields and finite element methods. *Geosci. Front.* **2018**, *9*, 1657–1664. [CrossRef]
7. Qiao, F.; Bo, J.; Qi, W.; Wang, L.; Chang, C.; Zhang, Z.; Wang, J. Study on the dynamic characteristics of soft soil. *RSC Adv.* **2020**, *10*, 4630–4639. [CrossRef]
8. Heede, P.V.D.; Belie, N.D. Environmental impact and life cycle assessment (LCA) of traditional and 'green' concretes: Liter-ature review and theoretical calculations. *Cem. Concr. Compos.* **2012**, *34*, 431–442. [CrossRef]
9. Lin, D.F.; Luo, H.L.; Hsiao, D.H.; Chen, C.T.; Cai, M.D. Enhancing soft subgrade soil with a sewage sludge ash/cement mixture and nano-silicon dioxide. *Environ. Earth Sci.* **2016**, *75*, 619. [CrossRef]
10. Ghasabkolaei, N.; Janalizadeh, A.; Jahanshahi, M.; Roshan, N.; Ghasemi, S.E. Physical and geotechnical properties of cement-treated clayey soil using silica nanoparticles: An experimental study. *Eur. Phys. J. Plus* **2016**, *131*, 134. [CrossRef]
11. Jongpradist, P.; Jumlongrach, N.; Youwai, S.; Chucheepsakul, S. Influence of Fly Ash on Unconfined Compressive Strength of Cement-Admixed Clay at High Water Content. *J. Mater. Civ. Eng.* **2010**, *22*, 49–58. [CrossRef]
12. Kong, D.; Du, X.; Wei, S.; Zhang, H.; Yang, Y.; Shah, S.P. Influence of nano-silica agglomeration on microstructure and properties of the hardened cement-based materials. *Constr. Build. Mater.* **2012**, *37*, 707–715. [CrossRef]
13. Mao, J.; Wu, Q.; Tao, F.; Xu, W.; Hong, T.; Dong, Y. Facile fabrication of porous BiVO4hollow spheres with improved visible-light photocatalytic properties. *RSC Adv.* **2020**, *10*, 6395–6404. [CrossRef]
14. Land, G.; Stephan, D. The influence of nano-silica on the hydration of ordinary Portland cement. *J. Mater. Sci.* **2012**, *47*, 1011–1017. [CrossRef]
15. Hou, P.; Qian, J.; Cheng, X.; Shah, S.P. Effects of the pozzolanic reactivity of nano SiO_2 on cement-based materials. *Cem. Concr. Compos.* **2015**, *55*, 250–258. [CrossRef]
16. Farzadnia, N.; Abang, A.A.A.; Demirboga, R.; Parvez, A.M. Effect of halloysite nanoclay on mechanical properties, thermal behavior and microstructure of cement mortars. *Cem. Concr. Res.* **2013**, *48*, 97–104. [CrossRef]
17. Wang, W.; Li, J.; Hu, J. Triaxial Mechanical Properties and Micromechanism of Calcareous Sand Modified by Nanoclay and Cement. *Geofluids* **2021**, *2021*, 1–9. [CrossRef]
18. Li, N.; Lv, S.; Wang, W.; Guo, J.; Jiang, P.; Liu, Y. Experimental investigations on the mechanical behavior of iron tailings powder with compound admixture of cement and nano-clay. *Constr. Build. Mater.* **2020**, *254*, 119259. [CrossRef]
19. Meyer, C. The greening of the concrete industry. *Cem. Concr. Compos.* **2009**, *31*, 601–605. [CrossRef]
20. Ruan, S.; Unluer, C. Comparative life cycle assessment of reactive MgO and Portland cement production. *J. Clean. Prod.* **2016**, *137*, 258–273. [CrossRef]
21. Gao, L.; Ren, Z.; Yu, X. Experimental Study of Nanometer Magnesium Oxide-Modified Clay. *Soil Mech. Found. Eng.* **2015**, *52*, 218–224. [CrossRef]
22. Gao, L.; Ren, K.Y.; Ren, Z.; Yu, X.J. Study on the shear property of nano-MgO-modified soil. *Mar. Georesour. Geotechnol.* **2018**, *36*, 465–470. [CrossRef]
23. Moradpour, R.; Taheri-Nassaj, E.; Parhizkar, T.; Ghodsian, M. The effects of nanoscale expansive agents on the mechanical properties of non-shrink cement-based composites: The influence of nano-MgO addition. *Compos. Part B* **2013**, *55*, 193–202. [CrossRef]
24. Wang, W.; Li, Y.; Yao, K.; Li, N.; Zhou, A.; Zhang, C. Strength properties of nano-MgO and cement stabilized coastal silty clay subjected to sulfuric acid attack. *Mar. Georesour. Geotechnol.* **2020**, *38*, 1177–1186. [CrossRef]
25. Yao, K.; Wang, W.; Li, N.; Zhang, C.; Wang, L.X. Investigation on strength and microstructure characteristics of Nano-MgO admixed with cemented soft soil. *Constr. Build. Mater.* **2019**, *206*, 160–168. [CrossRef]
26. Yao, K.; An, D.L.; Wang, W.; Li, N.; Zhang, C.; Zhou, A.Z. Effect of nano-MgO on mechanical performance of cement stabilized silty clay. *Mar. Georesour. Geotechnol.* **2020**, *38*, 250–255. [CrossRef]
27. Yuan, H.W.; Shi, Y.; Xu, Z.Z.; Lu, C.H.; Ni, Y.R.; Lan, X.H. Effect of nano-MgO on thermal and mechanical properties of alu-minate cement composite thermal energy storage materials. *Ceram. Int.* **2014**, *40*, 4811–4817. [CrossRef]
28. Hou, P.K.; Cai, Y.M.; Cheng, X.; Zhang, X.Z. Effects of the hydration reactivity of ultrafine magnesium oxide on cement-based materials. *Mag. Concr. Res.* **2017**, *69*, 1135–1145. [CrossRef]
29. Liu, J.H.; Hou, F.F.; Song, S.M.; Jia, B.Y. Effects of Compressive Strength and Carbonation of HCSA Amount on High Volume Fly Ash Concrete. *Adv. Mater. Res.* **2011**, *261–263*, 333–337. [CrossRef]
30. Cui, H.Z.; Tang, W.C.; Liu, W.; Dong, Z.J.; Xing, F. Experimental study on effects of CO_2 concentrations on concrete carbona-tion and diffusion mechanisms. *Constr. Build. Mater.* **2015**, *93*, 522–527. [CrossRef]
31. Yoon, I.-S.; Çopuroğlu, O.; Park, K.-B. Effect of global climatic change on carbonation progress of concrete. *Atmos. Environ.* **2007**, *41*, 7274–7285. [CrossRef]
32. Yi, Y.; Liska, M.; Unluer, C.; Al-Tabbaa, A. Carbonating magnesia for soil stabilization. *Can. Geotech. J.* **2013**, *50*, 899–905. [CrossRef]
33. Cai, G.H.; Liu, S.Y.; Du, Y.J.; Zhang, D.W.; Zheng, X. Strength and deformation characteristics of carbonated reactive magnesia treated silt soil. *J. Cent. South Univ.* **2015**, *22*, 1859–1868. [CrossRef]
34. Vandeperre, L.J.; Al-Tabbaa, A. Accelerated carbonation of reactive MgO cements. *Adv. Cem. Res.* **2007**, *19*, 67–79. [CrossRef]

35. Mo, L.; Panesar, D.K. Effects of accelerated carbonation on the microstructure of Portland cement pastes containing reactive MgO. *Cem. Concr. Res.* **2012**, *42*, 769–777. [CrossRef]
36. Unluer, C.; Al-Tabbaa, A. Impact of hydrated magnesium carbonate additives on the carbonation of reactive MgO cements. *Cem. Concr. Res.* **2013**, *54*, 87–97. [CrossRef]

sustainability MDPI

Article

Characteristics and Mechanical Properties of Graphene Nanoplatelets-Reinforced Epoxy Nanocomposites: Comparison of Different Dispersal Mechanisms

Ming-Yuan Shen *, Wen-Yuan Liao, Tan-Qi Wang and Wei-Min Lai

Department of Mechanical Engineering, National Chin-Yi University of Technology, Taichung 411, Taiwan; qwerty4731598@gmail.com (W.-Y.L.); p124178571@gmail.com (T.-Q.W.); willy890125@gmail.com (W.-M.L.)
* Correspondence: myshen@ncut.edu.tw; Tel.: +886-4-3924505 (ext. 7153)

Abstract: The preparation of polymer-based nanocomposites requires considerable time (i.e., the dispersal of nanomaterials into a polymer matrix), resulting in difficulties associated with their commercial use. In this study, two simple and efficient dispersion methods, namely planetary centrifugal mixing and three-roll milling, were used to enable the graphene nanoplatelets to disperse uniformly throughout an epoxy solution (i.e., 0, 0.1, 0.25, 0.5, and 1.0 wt.%) and allow the subsequent preparation of graphene nanoplatelets/epoxy nanocomposites. Measurements of mechanical properties of these nanocomposites, including ultimate tensile strength, flexural strength, and flexural modulus, were used to evaluate these dispersal methods. Dispersing graphene nanoplatelets into the epoxy resin by planetary centrifugal mixing not only required a shorter process time but also resulted in a more uniform dispersion of graphene nanoplatelets than that by three-roll milling. In addition, compared with traditional dispersal methods, planetary centrifugal mixing was a more efficient dispersal method for the preparation of epoxy-based nanocomposites.

Keywords: graphene nanoplatelets; graphene nanoplatelets/epoxy nanocomposite; mechanical properties; planetary centrifugal mixing; three-roll milling; traditional dispersal

Citation: Shen, M.-Y.; Liao, W.-Y.; Wang, T.-Q.; Lai, W.-M. Characteristics and Mechanical Properties of Graphene Nanoplatelets-Reinforced Epoxy Nanocomposites: Comparison of Different Dispersal Mechanisms. *Sustainability* **2021**, *13*, 1788. https://doi.org/10.3390/su13041788

Academic Editor: Yeou-Fong Li
Received: 31 December 2020
Accepted: 29 January 2021
Published: 7 February 2021

1. Introduction

Carbon atoms can be arranged in different ways to form different dimensions of nano-carbon materials, such as C60 (0D), carbon nanotubes (1D), graphene (2D), and nano-carbon aerogels (3D). Although they are all composed of carbon, their properties are very different. Figure 1 shows the different structures of nano-carbon materials [1].

The arrangement of carbon atoms in graphene is the same as that in the single layer of graphite, which is a single layer of two-dimensional crystals composed of carbon atoms arranged in a honeycomb crystal lattice with sp^2 mixed orbital domains. Graphene can be thought of as a lattice of atoms formed by carbon atoms and their covalent bonds. In general, a high specific surface area between a polymer and a nanoparticle maximizes the transfer of stress from the polymer matrix to the nanoparticle. This can effectively reduce the composite's cracking caused by the stress concentration. Furthermore, 2D nano-carbon materials such as graphene have a higher specific surface area than multi-walled carbon nanotubes (MWCNTs) and are not as easily entangled and difficult to disperse as MWCNTs. Therefore, graphene would be a better reinforcement for polymer composites than carbon nanotubes (CNTs). Graphene nanoplatelets (GNPs) are ultrathin stacks of graphene layers with more than 10 carbon layers and thicknesses in the range of 5–100 nm. In some literature, they are also referred to as "graphene nanosheets" [2–8] (Figure 2). However, the large surface area of GNPs produces large van der Waals forces and strong π–π interactions between the planar nanosheets [9–11]. Thus, the applications of GNP-reinforced polymer nanocomposites are limited because of the bad dispersion result in the aggregation of GNP.

Epoxy resin is a molecule containing two or more epoxide functional groups. The process of mixing the resin and the hardener at a certain ratio after the chemical reaction to form a three-dimensional network structure (cross-linking) is called curing. In addition, epoxy resins are widely used in advanced carbon fiber-reinforced plastics (CFRPs) owing to their good mechanical performance, processibility, compatibility with most fibers, chemical resistance, wear resistance, and low cost [12,13]. However, the higher cross-link densities of epoxy resins can contribute to low absolute strength and poor fracture toughness, thereby limiting the use of epoxy composites in applications involving mechanical components [14]. Various types of reinforcements have been developed to improve the mechanical properties of epoxy-based composites [15–18]. Hence, nano-carbon materials, such as CNT or GNP, are usually used to improve the mechanical properties of polymer composites and their fiber-reinforced composite laminates (CFRP). However, nano-carbon materials are difficult to disperse into polymer matrices, resulting in the agglomeration in the polymer, and thus have limited properties and applications. This is because the van der Waals forces between the carbon nanomaterials can lead to agglomeration and, consequently, require the use of complex processes to obtain a homogeneous dispersal in a polymer matrix [18–24].

In our previous studies [19,22], we demonstrated that adding GNPs into epoxy matrices through traditional methods of dispersal (TD) and the subsequent preparation of GNP/epoxy nanocomposites and their CFRP laminates could improve the mechanical and interlaminar properties of the nanocomposite and CFRP laminates, respectively. However, the TD processing time required to prepare nanocomposites was high, although it significantly improved the mechanical properties. This is because the agglomeration arising from the van der Waals forces of nano-carbon materials resulted in a lengthy process time for the dispersion of these materials into the polymer matrices.

In general, the factors affecting the dispersal of nano-scaled particles can be divided into three interacting phases (Figure 3): the stress application mechanism, the operating method, and the specific energy supply are decided by the dispersal equipment. In contrast, the particle properties specify the resistance against the fragmentation of aggregates in the dispersal process. Basically, this includes the material properties, the surface modification, particle size distribution, and the particle–particle interactions (i.e., the surface modification of particles). For nano-particulate agglomerates and aggregates, particle–particle interactions are especially important and are very strong; they can be described according to the Derjaguin–Landau–Verwey–Overbeek theory [25,26]. The third key point is formulation, including the liquid phase of the nanoparticles, rheological properties, ion concentration, and additives.

The use of a three-roll milling apparatus has recently produced good results for nano-reinforcement dispersal. This method exerts shear forces over the particles and avoids the presence of compression forces. During three-roll milling, the adjacent rolls rotate in opposite directions, and the external shear forces generated in the gap between the rolls break large agglomerates into smaller aggregates and further into primary particles [27]. Thus, agglomerates can be separated without being damaged. In several studies [28–35], the three-roll milling process has proven to be an effective method for producing GNP/epoxy nanocomposites with a homogenous dispersion.

Planetary centrifugal mixing (PCM) has several advantages over other types of mixing using blade agitators [36,37]. The PCM apparatus utilizes centrifugal swirling, which causes less or no damage to materials that would be otherwise damaged by agitating blades. It can effectively de-aerate the mixture by means of large centrifugal forces and prevent contamination from outside, because the vial is perfectly sealed and no power-transmitting shafts need to pass into the interior of the vial. Consequently, PCM can be used for the rapid and homogeneous blending and dispersal of organic and inorganic powders, rheological fluids, and their mixtures, including, but not limited to, nano-and micro-sized powders [38–44], magnetic powders [45], and nano- and micro-particle suspensions [46,47].

Although the dispersal effectiveness has been established in the aforementioned studies and there are several dispersal processes for each type of nano-reinforcement/resin mixture, there is a lack of in-depth knowledge regarding the differences in the effectiveness of the dispersion attained. Furthermore, very few studies have been undertaken on the fabrication of nano-carbon materials, such as GNP-reinforced polymer nanocomposites, using different methods of dispersal with a subsequent investigation of their mechanical properties. Therefore, there remains a need to investigate the dispersal and the mechanical properties of GNP-reinforced epoxy resin.

In this study, two simple and efficient methods of dispersal (planetary centrifugal mixing (PCM) and three-roll milling (TRM)), were used to enable a uniform dispersal of GNP throughout an epoxy solution (i.e., 0, 0.1, 0.25, 0.5, and 1.0 wt.%), with the subsequent preparation of GNP/epoxy nanocomposites. The mechanical properties of the nanocomposites, including ultimate tensile strength, flexural strength, and flexural modulus, were investigated. To understand the differences between the traditional process of dispersal and these two dispersal processes, the results of our previous study [19] were compared to those of the current study.

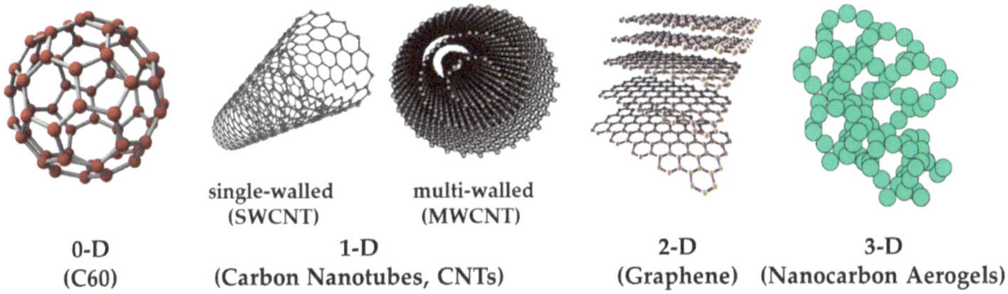

Figure 1. Different structures of nano-carbon materials.

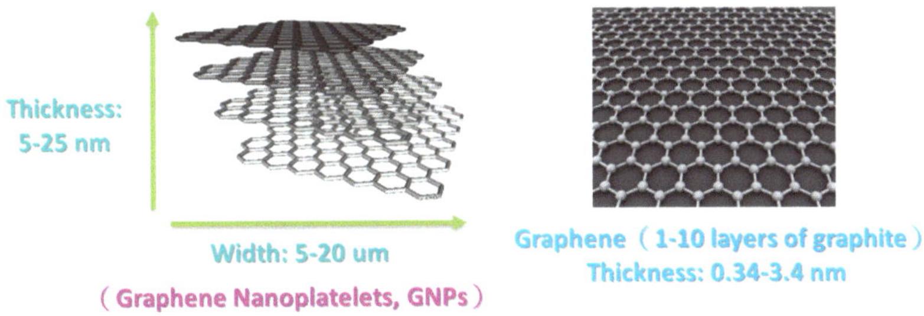

Figure 2. Graphene and graphene nanoplatelets (GNPs) [48].

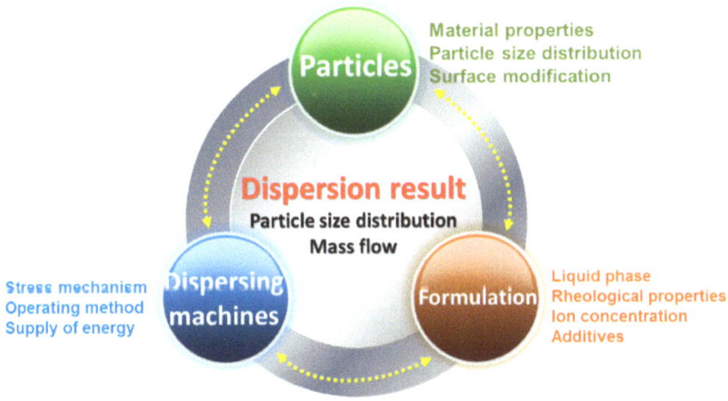

Figure 3. Factors affecting dispersal [26].

2. Materials and Methods

2.1. Nano-Carbon Materials

The GNP diameter was 3–5 μm, the thickness was approximately 5–25 nm, and the purity of the GNPs was >98%. Graphene is a single atomic layer of graphite with a thickness of 0.34 nm. In the industrial field, a single layer of graphite with less than 10 layers can be commonly referred to as "graphene"; the thickness of graphite can be several hundred microns. The nano-carbon material used in this study had a thickness of approximately 5–25 nm (within 100 layers), which was different from the structure of graphene and graphite. Therefore, this material was referred to as graphene nanoplatelets. The aspect ratio of GNP was approximately 30–60 m^2/g, which enhanced the contact area with the polymer matrix. The theory bulk density of the GNP was 1.6 g/cm^3. A transmission electron microscope (TEM) was used to determine the aspect ratio, and a field-emission scanning electron microscope (FESEM) was used to determine the thickness. The images obtained revealed that the surface morphology of GNP was considerably different to that of carbon black, CNTs, and graphite (Figure 4). The GNP particles were aggregated by van der Waals forces.

The carbon structure of carbon black differed from that of graphite and graphene or even GNPs in that it consisted of concentrically aligned vortex layers of graphite domains. This structure was proposed by Heidenreich et al. in 1968 [49], as shown in Figure 5.

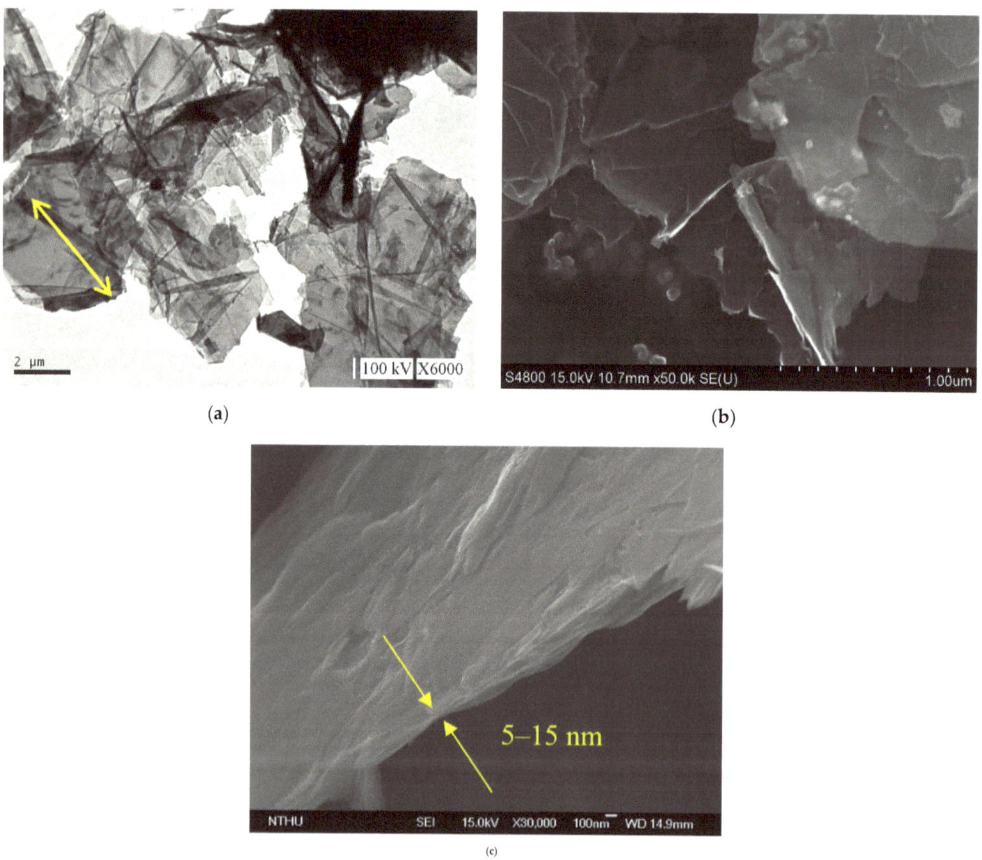

Figure 4. Morphologies of GNP: (**a**) transmission electron microscope (TEM) image, (**b**) scanning electron microscope (SEM) image of the GNP ×10,000, and (**c**) ×30,000.

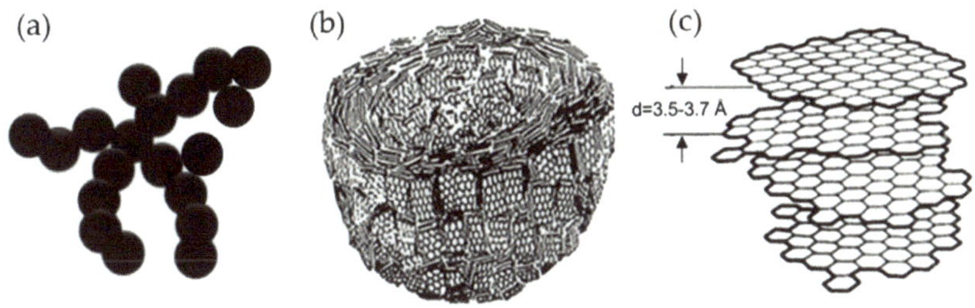

Figure 5. Carbon structure of carbon black: (**a**) macroscopic for carbon black aggregation, (**b**) aligned vortex layers of graphite domains, and (**c**) distance between carbon layers of carbon black [49].

2.2. Preparation of GNP/Epoxy Resin Nanocomposites

Two GNP/epoxy solutions were prepared using two different mixing techniques. One was a planetary centrifugal mixer (Mazerustar KK-250S; Kurabo Industries Ltd., Osaka, Japan), and the other was a three-roll miller (RY-R065; Renyih Industrial Co., Ltd., Taiwan).

2.2.1. Preparation of GNP/Epoxy Resin Solution Using Planetary Centrifugal Mixing

The GNPs obtained from Xiamen Knano Graphene Technology Co., Ltd., China, were first dried at 120 °C for 120 min in order to eliminate the moisture within them. A methyl ethyl ketone (MEK) solvent was mixed with GNP and then vibrated by ultrasonication for 30 min to enable a homogeneous GNP dispersion. A solvent-type epoxy resin (EPO-622™ obtained from Epotech Composite Co., Ltd., Taiwan) was then added into the MEK/GNP solution, and the mixture was stirred for 20 min by using a planetary centrifugal mixing apparatus to enable the MEK/GNP to disperse uniformly throughout the epoxy resin. This mixer used a mechanism whereby the container holding the material revolved clockwise while the container itself rotated counter-clockwise (rotation). Finally, the MEK/GNP/epoxy solution was placed in an oven at 83 °C to evaporate the solvent (MEK). A schematic representation of the fabrication of the GNP/epoxy solution by PCM is shown in Figure 6a.

The planetary centrifugal mixing machine consisted of a cylindrical container that generally had a flat bottom with no obstacles or agitators. The container was subject to two rotational movements: one was rotation around its geometric axis, and the other was a revolution around an axis located at a given distance (R_0). Figure 7 presents a schematic representation of this instrument, which approximated the shape and the function of a standard centrifuge. The difference from a centrifuge was that a plastic container was also rotated at an appropriate speed. The mixing container rotated in the opposite direction to the central rotating shaft at speed, as well as the central rotating drum, thus generating high shear forces in the container. This combination homogenized the nanoparticle in the epoxy matrix. Using this mixer, the sub-micrometer-sized (diameter) GNP particles could be dispersed homogeneously into the resin within 20 min, which was a shorter treatment time than that obtained in our previous studies [19,22].

2.2.2. Preparation of GNP/Epoxy Resin Solution by Three-Roll Milling (TRM)

The GNPs were dried in an oven at 120 °C for 120 min to eliminate the moisture within them. They were subsequently mixed with epoxy resin for 30 min by using a mechanical mixer and then mixed using a three-roll milling apparatus for 100 min to improve the uniformity of the GNP distribution in the epoxy solution. The gap size between each pair of rolls varied as a function of the steps (in microns): (1) 320–340; (2) 260–290; (3) 210–230; (4) 160–180; and (5) 110–130. The dispersion time of one gap was 20 min, and the total time of all the five gaps was 100 min. During the three-roll milling process, rollers 1 and 3 rotated in the same direction, whereas roller 2, located between 1 and 3, rotated in the opposite direction, thereby inducing high shearing in the mixture. The three rollers rotated at different speeds, with the speed ratio among them maintained at 9:3:1 ($\omega_3 = 3\omega_2 = 9\omega_1$) in each of the three passes. The mismatch between the angular velocity of the adjacent rolls and the direction of rotation (clockwise or anticlockwise) was controlled to exert high shear forces over the mixture (Figure 8a). The mixture was poured between the two first rolls and was collected after the third roll. The separation between the rolls influenced the forces exerted over the mixture and therefore on the dispersion achieved. Smaller distances between the rolls produced greater shear forces, which were necessary to break large agglomerates. The reduction of the agglomeration size resulted in a better dispersion of the nano-reinforcement. Nevertheless, when the force exerted was too large, it might break the structure of GNPs, leading to a reduced aspect ratio of the GNPs. The principle of the flow conditions can be seen in Figure 8b [50]. A primary dispersion of the agglomerates could be achieved by the knead-vortex between the rolls, while the final exfoliation and dispersion of the GNPs occurred in the area between the rolls. During the dispersion process, the adjacent rollers rotated at different speeds and in opposite directions. The dispersion effect resulted from the high shear stresses generated in the gap between the rolls. A schematic representation of the fabrication of the GNP/epoxy solution by the three-roll milling process is shown in Figure 6b.

2.2.3. Preparation of GNP/Epoxy Nanocomposites

The GNP/epoxy resin solution (dispersed through one of two processes) was first placed in a vacuum heating oven to eliminate air bubbles and residual solvents. The resin solution was then poured into molds and placed on a hot-press machine to prepare the GNP/epoxy nanocomposites. A schematic representation of the fabrication of the GNP/epoxy nanocomposite is shown in Figure 6c. The curing cycle of the nanocomposite is shown in Figure 9. Note that the temperature of the epoxy solution was raised from room temperature to 120 °C and held for 5 min to increase the flow of the resin as well as to allow the degassing from the mold. The temperature was then raised to 150 °C; the resin started to enter curing, and the pressurization process started at this time. In this period, the perfect composite could be prepared under the effect both temperature and pressure. After holding the temperature for 30 min, the composite started to cool; it took approximately 240 min for the composite to cool from 150 °C to room temperature. Finally, the nanocomposite could be demolded. After the preparation of the nanocomposites was completed, the nanocomposites were placed in a heating oven at 120 °C for 120 min to eliminate the internal stress; this was called the "post-curing" process.

2.3. *Test Methods*

2.3.1. Tensile Test

The tensile strength of GNP/epoxy nanocomposites was measured using a universal materials testing system (Instron 3369). The tensile testing specimens with a dumb-bell shape were prepared by machining samples from 3-mm-thick plates; the samples had a gauge length of 50 mm, according to ASTM D 638 type I. An extensometer with a 50-mm gauge length was used for measuring the axial strain and was stuck on the surfaces. During the tests, the specimens were subjected to tensile loading at a crosshead speed of 1 mm min^{-1} at room temperature. The dimensions of the tensile specimen are shown in Figure 10a. Five specimens of each nanocomposite were taken and tested for the ultimate tensile strength σ_t, which was obtained from the average value as follows:

$$\sigma_t = \frac{P}{A}$$

where P is the maximum tension loading (N) and A is the section area.

2.3.2. Flexural Test

The flexural properties were determined according ASTM D790-17 by using the three-point bending test method. The dimension of the rectangular flexural specimen was

130 mm × 13 mm × 3 mm, and the span was 16 times the thickness according to the standard, i.e., 48 mm. As the thickness of the specimen varied slightly because of the preparation process, the average thickness of the specimen was used as the span setting value before each test, and then the span and drop rate were calculated according to this value. The bending test value was obtained by averaging the test data. The maximum loads were obtained and the flexural strength (σ_f) and modulus (E_B) were calculated using the following formulas respectively; the data were obtained from the average of the five tested values. The dimensions of flexural specimen are shown in Figure 10b.

$$\sigma_f = \frac{3PL}{2Wt^2}$$

$$E_B = \frac{L^3 m}{4Wt^3}$$

where P is the maximum load (N); L is the span, which is the distance between the supports (mm); m is the initial slope of the linear region of the load/displacement curve; W is the width of the specimen (mm); and t is the thickness (mm).

2.3.3. Viscosity Test

In this study, the kinetic viscosity was calculated according to the kinematic viscosity measurement described in the ASTMD445 standard: η in the fluid with two areas of 1 m^2, 1 m apart, and a relative speed of movement of 1 m/s; the resistance thus faced is called kinetic viscosity. The unit of kinetic viscosity is pascal-seconds. The kinetic viscosity η was calculated by using the following formula:

$$\eta = \varrho \upsilon$$

where η is the kinetic viscosity; ϱ is the density of the fluid, and υ is the kinematic viscosity.

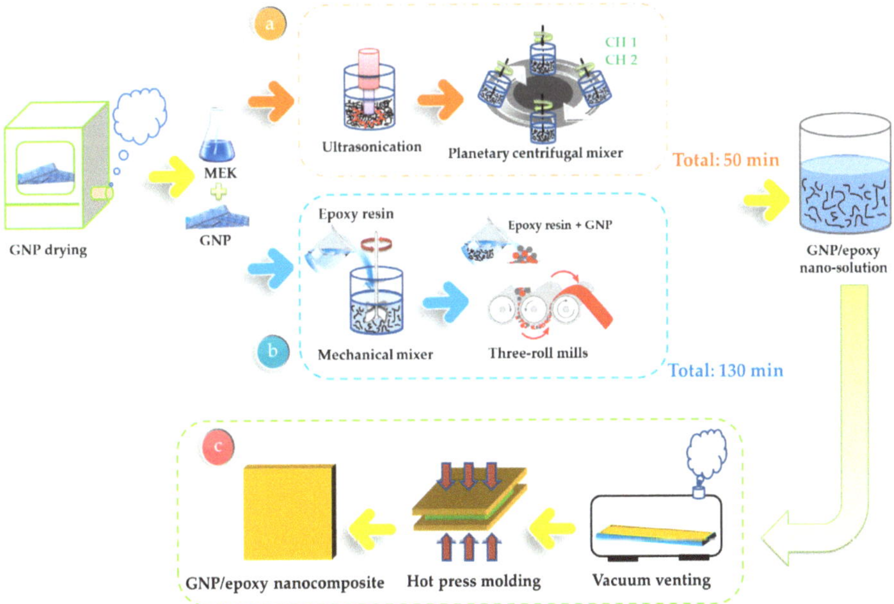

Figure 6. Schematic representation of the preparation of (**a**) GNP/epoxy solution through the planetary centrifugal mixing process, (**b**) GNP/epoxy solution through the three-roll milling process, and (**c**) GNP/epoxy nanocomposites.

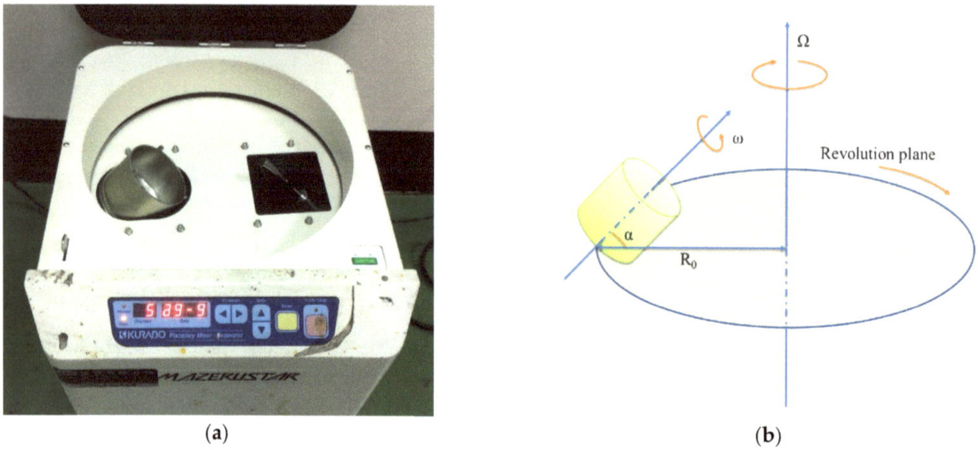

Figure 7. Schematic representation of the planetary centrifugal mixer: (**a**) Kurabo Mazerustar KK-250S and (**b**) schematic representation of this instrument.

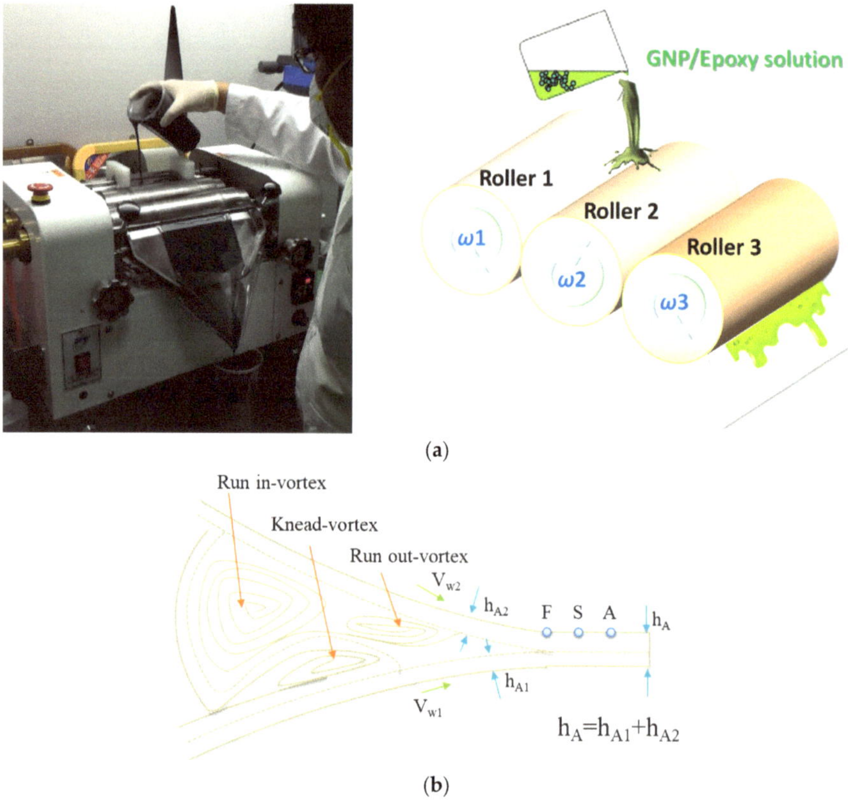

Figure 8. (**a**) Three-roll milling used for the dispersion of GNPs in an epoxy matrix, and (**b**) schematic representation of the flow conditions in the roller clearance [50].

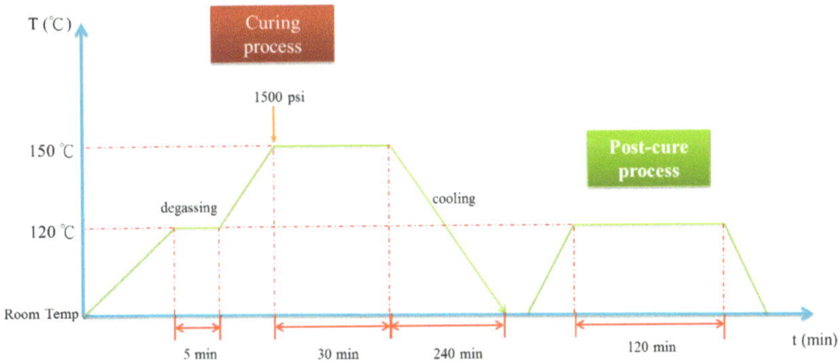

Figure 9. Curing cycle of nanocomposite.

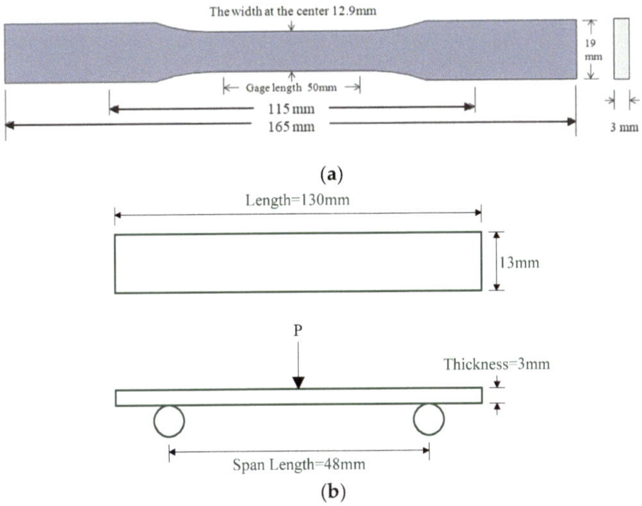

Figure 10. Dimensions of nanocomposite specimen: (**a**) tensile and (**b**) flexural specimen.

3. Results and Discussion

3.1. Mechanical Properties of GNP/Epoxy Nanocomposites Prepared Using Planetary Centrifugal Mixing

The planetary centrifugal mixing equipment used in this study enables simultaneous mixing/de-aerating within a short time without the use of mixing rods and blades. In order to find the best planetary centrifugal mixing condition, two different parameters were used to disperse GNPs into epoxy, followed by an investigation of the mechanical properties of the GNP/epoxy nanocomposite.

Table 1 summarizes the different dispersal parameters of the PCM used in this study. The rate of revolution could be varied between approximately 640 and 1700 rpm, with 10 increments. The rate of rotation also had 10 increments and ranged between 0.0 and 1.0 times the speed of revolution. The revolution speed of level 10 was approximately 1700 rpm. The revolution and the rotation speed of Channel 1 (CH 1) was approximately 1200 rpm and approximately 500 rpm, respectively. The equivalent values for CH 2 were approximately 1600 and 1600 rpm, respectively. CH 1 is predominantly used for stand mixing and de-aerating, and CH 2 is focused on mixing. In this study, these two different mixing channels were used to investigate the efficiency of GNP dispersal into the epoxy resin because the impact of de-aeration on the dispersal of GNPs into the epoxy was not known.

Figures 11–13 and Table 2 show the ultimate tensile strengths, flexural strengths, and flexural moduli of the epoxy nanocomposites with different contents of GNPs. The epoxy resin containing GNPs prepared with either channel of the PCM exhibited different mechanical properties as compared to the pure resin. The dispersion experienced with CH 1 emphasized standard dispersion (including mixing, degassing, and temperature suppression), which differed from CH 2′s emphasis on mixing. Therefore, when GNPs were mixed with the epoxy resin, the poor dispersion resulted in the poor mechanical properties of the nanocomposites.

When GNPs were combined with epoxy through the CH 2 dispersal, the mechanical properties of the nanocomposite with GNPs added at 0.25 wt.% exhibited the best enhancement, as compared to the neat epoxy resin without any GNPs added. In our previous study [19], the GNPs were used to reinforce the epoxy composite to enhance their mechanical properties using the traditional dispersal (TD) process. This earlier study also demonstrated that when GNPs were added at 0.25 wt.%, the mechanical properties of nanocomposites were enhanced the most. To investigate the effect of different dispersal mechanisms on the mechanical properties of epoxy nanocomposites reinforced with GNP, CH 2 was used in this study to prepare a GNP/epoxy nanocomposite; this was compared with the results obtained using other mixing processes.

Table 1. Operating parameters for the different applications of planetary centrifugal mixing (PCM).

Channel	Step			Function				Application
	Rev (rpm)	Rot (rpm)	Time (min)	Mixing	Deaerating	Speed	Temperature-Rise Control	
CH 1	1200	500	20	○	○	○	○	Standard mixing/deaerating
CH 2	1600	1600	20	◎	×	◎	×	Mixing priority

※ Rev: Revolution level; Rot: Rotation level; Time: Process time; ◎: Excellent; ○: Good; ×: Poor.

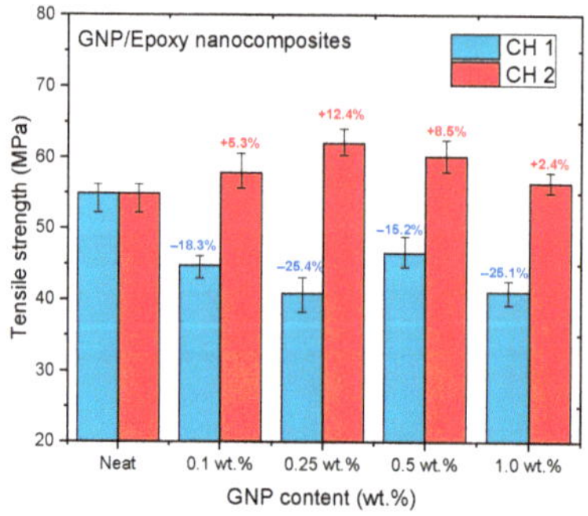

Figure 11. Tensile strengths of GNP/epoxy nanocomposites prepared using PCM.

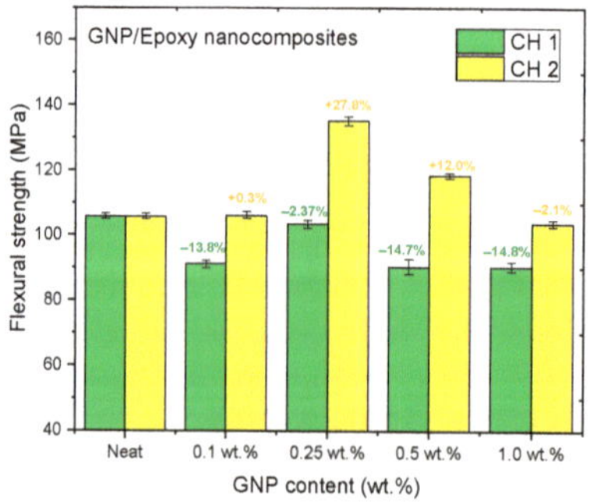

Figure 12. Flexural strengths of GNP/epoxy nanocomposites prepared using PCM.

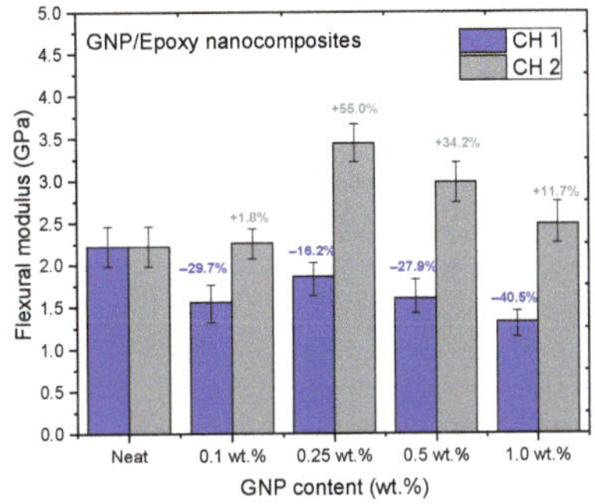

Figure 13. Flexural moduli of GNP/epoxy nanocomposites prepared using PCM.

Table 2. Mechanical properties of GNP/epoxy nanocomposites under different PCM conditions.

Unit	Channel	GNP Content (wt.%)				
		Neat	0.1	0.25	0.5	1.0
Tensile strength (MPa)	CH 1	54.89	50.82	53.87	60.21	53.02
	CH 2		44.82	40.93	46.57	41.14
Flexural strength (MPa)	CH 1	105.89	105.00	107.67	122.57	124.56
	CH 2		109.29	92.77	82.89	103.91
Flexural modulus (GPa)	CH 1	2.22	2.30	2.82	2.94	3.19
	CH 2		1.56	1.86	1.60	1.32

3.2. Mechanical Properties of GNP/Epoxy Nanocomposites Prepared Using Different Methods of Dispersal

Figures 14–16 and Tables 3–5 show the tensile strengths, flexural strengths, and flexural moduli of the epoxy nanocomposites fabricated with different GNP contents, using three dispersing processes. The mechanical properties of the nanocomposites prepared using PCM and TRM were investigated in this study, and the data pertaining to the traditional dispersal (TD) was taken from our previous study [22].

Nanocomposites prepared using PCM exhibited the same reinforcement effect as that observed with TD, and both exhibited better mechanical properties than pure resin. In our previous study, the use of TD to prepare nanocomposites demonstrated that the mechanical properties of GNP/epoxy nanocomposites were optimal when 0.25 wt.% GNP was added. These results were similar to those obtained using PCM. Furthermore, Figures 15 and 16 indicate that the flexural strengths and the flexural moduli of the nanocomposites without the added GNPs were approximately 105.89 MPa and 2.22 GPa, respectively. The reinforced nanocomposite containing 0.25 wt.% GNP, generated using PCM, exhibited very significantly improved flexural strength (27.8%) and flexural modulus (55.5%) compared to the neat epoxy resin.

The nanocomposites showed very significant mechanical enhancement by PCM and TD; Table 6 compares the different durations of dispersal with GNPs mixed into epoxy. For the TD process, although significant improvements in the mechanical properties (from 0.25–1 wt.%) were apparent, the time required for processing was reasonably long (a total of 190 min). In contrast, the TRM processing time was shorter than that required in the TD process. The epoxy resin used in this study was a solvent-type (low-viscosity) resin, and the open operating space resulted in solvent evaporation, thereby causing a variation in the viscosity within the different gaps applied by the TRM process (Figure 8a). Resin viscosity was the key factor in the dispersal of nanomaterials into the resin. To ensure that GNPs could be effectively dispersed into the epoxy resin by three-roll milling, the solvent (MEK) had to be irregularly added to reduce the variability in the viscosity of the resin during TRM. Consequently, the viscosity of the resin changed irregularly during TRM. The viscosities of EPO-622^{TM} and GNP/epoxy (TRM) at 30 °C are presented in Figures 17 and 18 and Table 7. The viscosity of the GNP/epoxy resin produced by TRM exhibited significant changes, which resulted in the instability and non-uniformity of the dispersal of GNPs into epoxy resin and in turn the instability of the mechanical properties of nanocomposites.

The high speed and shorter processing time for dispersal required by PCM aided in maintaining stability and resulted in a significant improvement of the mechanical properties of nanocomposites. This was because of the high shear strength imposed in a close dispersing space. The mechanism of rotation in PCM was similar to the vortex fluidic exfoliation process [51]. The aggregation and stacking between GNP particles arose because of the van der Waals forces. In the PCM device, the high internal shear stresses within the tube wall provided the necessary conditions for dispersal. According to fluid dynamics, the liquid flow was upwards at the internal surface of the rotating tube, and downwards close to the liquid surface (Stewartson/Ekman layers) (Figure 19a). The shear layers were parallel to the axis of rotation in the rapidly rotating fluid, and the GNP particles initially accelerated to the walls of the tube because of the large centrifugal force.

In the rapidly rotating fluid, the shear layers were parallel to the axis of rotation. The large centrifugal force initially accelerated the GNPs to the walls of the tube, as shown in Figure 19b,c; the GNP particle exfoliation occurred via this rotation process. This required the lifting of individual particles from the surface of the bulk GNP, arising through the presence of a lateral force. The exfoliation involved a slippage process, whereby the GNPs were displaced relative to each other. Therefore, a large centrifugal force ensured that the GNPs experienced a shear-induced displacement along the surface.

Planetary centrifugal mixing simultaneously combined rotation and revolution. A powerful centrifugal force enabled the materials to be well dispersed. Rotation could effectively exfoliate GNPs, while revolution could further disperse GNPs into the resin and complete the de-aeration. Thus, CH 2, with a processing time of 20 min, was found to be the optimal setting for dispersing GNPs into epoxy resin in this study.

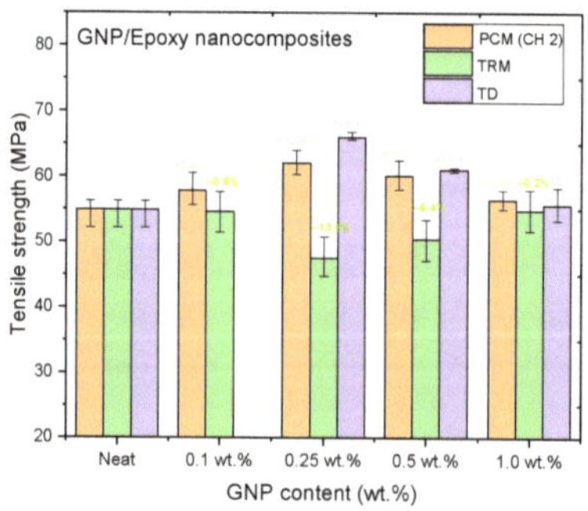

Figure 14. Tensile strengths of GNP/epoxy nanocomposites prepared using different dispersing processes.

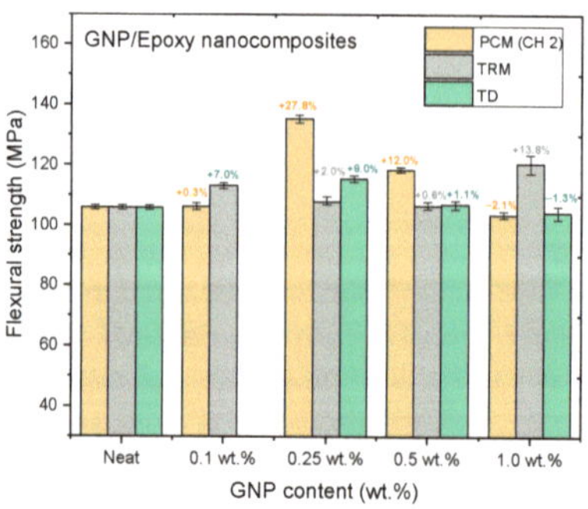

Figure 15. Flexural strengths of GNP/epoxy nanocomposites prepared using different dispersing processes.

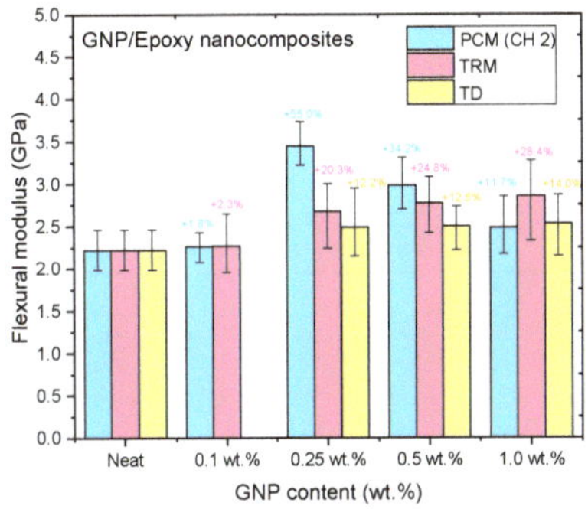

Figure 16. Flexural moduli of GNP/epoxy nanocomposites prepared using different dispersing processes.

Table 3. Tensile strengths of GNP/epoxy nanocomposites prepared using different dispersing processes.

Process (MPa)	GNP Content (wt.%)				
	Neat	0.1	0.25	0.5	1.0
PCM (CH 2)		57.87	62.05	60.17	56.36
% Enhancement		*+5.4*	*+12.4*	*+8.5*	*+2.44*
TRM	54.89	54.57	47.46	50.41	54.79
% Enhancement		*−0.6*	*−13.6*	*−9.4*	*−0.2*
TM		-	66	61.11	55.59
% Enhancement		-	*+20.2*	*+11.3*	*+1.3*

※ GNP/epoxy nanocomposites: PCM: planetary centrifugal mixing process, TRM: three-roll milling process, TM: traditional mixing process.

Table 4. Flexural strengths of GNP/epoxy nanocomposites prepared using different dispersing g processes.

Process (MPa)	GNP Content (wt.%)				
	Neat	0.1	0.25	0.5	1.0
PCM (CH 2)		106.21	135.35	118.57	103.66
% Enhancement		*+0.3*	*+27.8*	*+12.0*	*−2.1*
TRM	105.89	113.27	108	106.55	120.54
% Enhancement		*+7.0*	*+2.0*	*+0.6*	*+13.8*
TM		-	115.46	107.07	104.48
% Enhancement		-	*+9.0*	*+11.3*	*+1.3*

Table 5. Flexural modulus of GNP/epoxy nanocomposites prepared using different dispersing processes.

Process (GPa)	GNP Content (wt.%)				
	Neat	0.1	0.25	0.5	1.0
PCM (CH 2)		2.26	3.44	2.98	2.48
% Enhancement		*+1.8*	*+55.0*	*+34.2*	*+11.7*
TRM	2.22	2.27	2.67	2.77	2.85
% Enhancement		*+2.3*	*+20.3*	*+24.8*	*+28.4*
TM		-	2.49	2.50	2.53
% Enhancement		-	*+12.2*	*+12.6*	*+14.0*

Table 6. Different process times for dispersion of GNP/epoxy solution.

Processes (Min)	Homogenizer Stirring	Ultrasonication	Mechanical Mixer	Planetary Centrifugal Mixer	Three-Roll Milling	Total Dispersion Time
PCM		30		20		50
TRM			30		100	130
TD	10	90	90			190

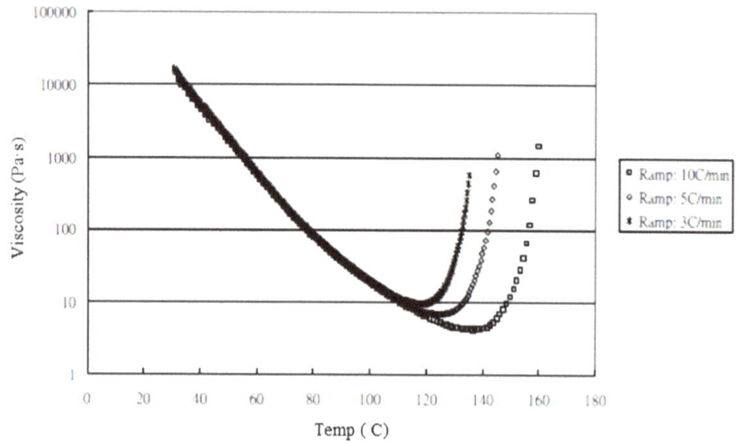

Figure 17. Viscosity curves of epoxy resin (EPO-622^{TM}).

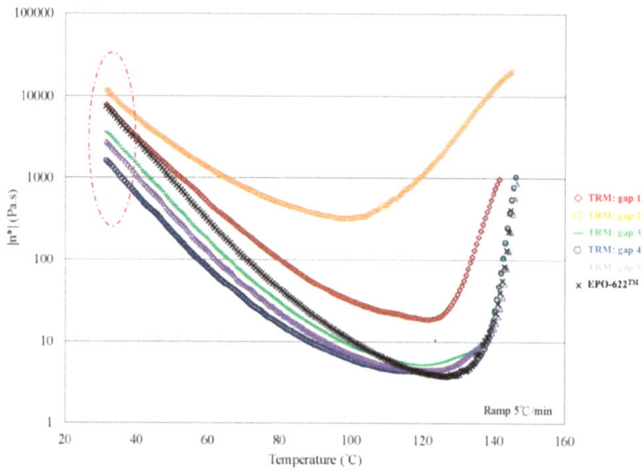

Figure 18. Viscosity curves of GNP/epoxy produced by TRM.

Table 7. Variable viscosity of epoxy resin under different TRM gaps at 30 °C.

Ramp 5 °C/min	EPO-622^{TM}	TRM				
		Gap 1	Gap 2	Gap 3	Gap 4	Gap 5
@30 °C, Viscosity (Pa·s)	8.6×10^3	9.2×10^3	1.4×10^4	4.2×10^3	1.9×10^3	3.1×10^3

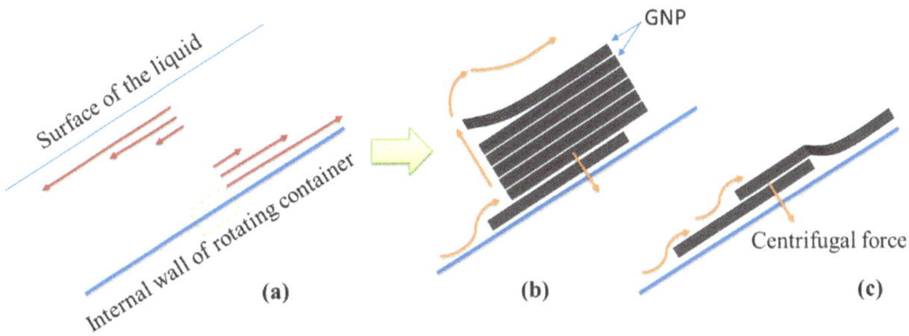

Figure 19. Schematic representation of (**a**) microfluidic flow velocity indicted by red arrows for a section of the rotating container, (**b**) dispersion process, and (**c**) slippage on the inner surface of the container.

3.3. Morphology of GNP/Epoxy Nanocomposites

In order to confirm the dispersal mechanism of the GNP-reinforced epoxy nanocomposite under two different dispersing processes, the FESEM was used to identify the fracture surface of the nanocomposite added with 0.25-wt.% GNPs. Figure 20a shows an scanning electron microscope (SEM) image of a neat cured-epoxy resin without GNP added; here, it can be seen that the breaking surface of the resin was quite smooth. Figure 20b,c show the broken section of the GNP/epoxy nanocomposite with the 0.25-wt.% GNP addition after the tensile test. With the same amount of GNP added, the average particle size of GNPs was smaller and the distribution in the resin was more uniform for the PCM dispersal process. This indicated that the shear force of the PCM method was stronger and could effectively overcome the van der Waals force between the particles, resulting in a better dispersion effect. Moreover, when the GNPs had good dispersion, the GNPs located in the crevices in the corrugation area restrained the crevice growth. Therefore, the mechanical properties of the nanocomposite could be significantly improved.

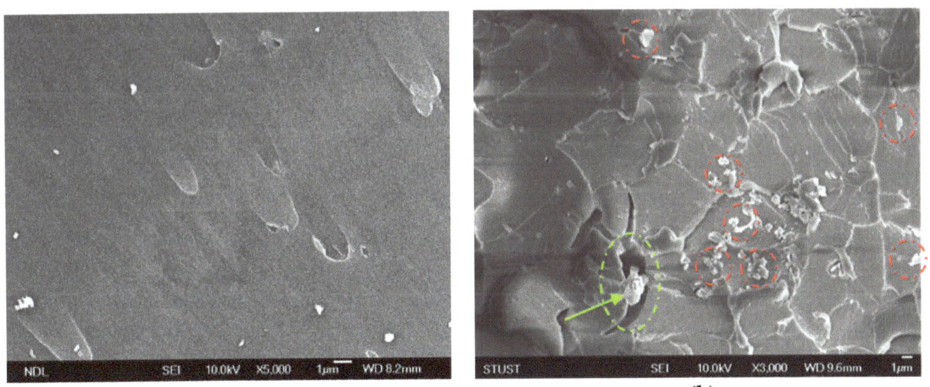

(**a**) (**b**)

Figure 20. *Cont.*

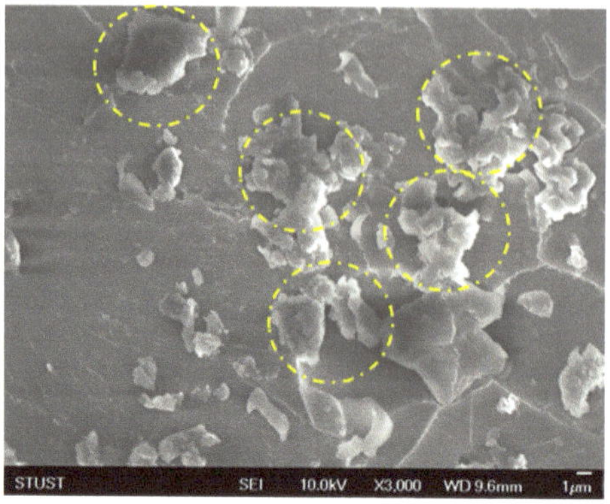

(c)

Figure 20. SEM images of nanocomposites prepared using different dispersing processes: (**a**) neat cured-epoxy, (**b**) 0.25-wt.% GNP dispersed by PCM process, and (**c**) 0.25-wt.% GNPs dispersed by TRM process.

4. Conclusions

All the three methods of dispersal improved the mechanical properties of the GNP/epoxy nanocomposites. For the TD process, the poor efficiency arose from the longer processing time. Although TRM improved the mechanical properties of the nanocomposites, the variable viscosity of the GNP/epoxy solution during the dispersal resulted in the unstable mechanical properties of the nanocomposites.

The morphological analysis by SEM revealed that the average particle size of the GNPs dispersed by the PCM method was not only smaller but also more uniformly dispersed around the resin, which was the key reason for the better mechanical properties. Therefore, the dispersion of GNPs into epoxy resin by the PCM method was not only efficient in terms of the process time but also significantly improved the mechanical properties of the nanocomposites. PCM could thus be deemed a more efficient and economically viable method of dispersal for the preparation of epoxy-based nanocomposites.

Author Contributions: Conceptualization, M.-Y.S.; methodology, M.-Y.S.; validation, W.-Y.L.; formal analysis, T.-Q.W.; investigation, M.-Y.S., W.-Y.L. and T.-Q.W.; resources, M.-Y.S.; data curation, W.-M.L.; writing—original draft preparation, M.-Y.S.; visualization, M.-Y.S.; supervision, M.-Y.S.; project administration, M.-Y.S.; funding acquisition, M.-Y.S. All authors have read and agreed to the published version of the manuscript.

Funding: This study is supported by the Ministry of Science and Technology (MOST) of Taiwan to National Chin-Yi University of Technology under a Contract No. MOST 107-2221-E-167-030-MY2.

Institutional Review Board Statement: Not applicable.

Informed Consent Statement: Not applicable.

Data Availability Statement: Not applicable.

Acknowledgments: This study was supported by the Ministry of Science and Technology (MOST) of Taiwan to National Chin-Yi University of Technology under MOST 107-2221-E-167-030-MY2. The authors also thank the Department of Biomechatronic Engineering, National Chiayi University, Taiwan, for guidance on analysis principles, and Epotech Composite Co., Ltd., Taiwan, who help with resin viscosity analysis was invaluable.

Conflicts of Interest: The authors declare no conflict of interest.

References

1. Sharma, V.K.; McDonald, T.J.; Kim, H.; Garg, V.K. Magnetic graphene-carbon nanotube iron nanocomposites as adsorbents and antibacterial agents for water purification. *Adv. Colloid Interface Sci.* **2015**, *225*, 229–240. [CrossRef]
2. Li, J.; Kim, J.K. Percolation threshold of conducting polymer composites containing 3D randomly distributed graphite nanoplatelets. *Compos. Sci. Technol.* **2007**, *67*, 2114–2120. [CrossRef]
3. Weng, W.G.; Chen, G.H.; Wu, D.J.; Yan, W.L. HDPE/expanded graphite electrically conducting composite. *Compos. Interf.* **2004**, *11*, 131–143. [CrossRef]
4. Chen, G.H.; Wu, D.J.; Weng, W.G.; Yan, W.L. Preparation of polymer/graphite conducting nanocomposite by intercalation polymerization. *J. Appl. Polym. Sci.* **2001**, *82*, 2506–2513. [CrossRef]
5. Zheng, W.; Wong, S.C. Electrical conductivity and dielectric properties of PMMA/expanded graphite composites. *Compos. Sci. Technol.* **2003**, *63*, 225–235. [CrossRef]
6. Chen, X.M.; Shen, J.W.; Huang, W.Y. Novel electrically conductive polypropylene/graphite nanocomposites. *J. Mater. Sci. Lett.* **2002**, *21*, 213–214. [CrossRef]
7. Yao, H.; Hawkins, S.A.; Sue, H.J. Preparation of epoxy nanocomposites containing well-dispersed graphene nanosheets. *Compos. Sci. Technol.* **2017**, *146*, 161–168. [CrossRef]
8. Li, A.; Zhang, C.; Zhang, Y.F. Graphene nanosheets-filled epoxy composites prepared by a fast dispersion method. *J. Appl. Polym. Sci.* **2017**, *134*, 45152. [CrossRef]
9. Si, Y.; Samulski, E.T. Exfoliated graphene separated by platinum nanoparticles. *Chem. Mater.* **2008**, *20*, 6792–6797. [CrossRef]
10. Si, Y.; Samulski, E.T. Synthesis of water soluble graphene. *Nano. Lett.* **2008**, *8*, 1679–1682. [CrossRef] [PubMed]
11. Li, D.; Muller, M.B.; Gilje, S.; Kaner, R.B.; Wallace, G.G. Processable aqueous dispersions of graphene nanosheets. *Nat. Nanotechnol.* **2008**, *3*, 101–105. [CrossRef]
12. Ram, A. *Fundamentals of Polymer Engineering*; Springer: New York, NY, USA, 1997.
13. Kinloch, A.J. *Adhesion and Adhesives: Science and Technology*, 1st ed.; Chapman and Hall: London, UK, 1987.
14. Kinloch, A.J.; Lee, S.H.; Taylor, A.C. Improving the fracture toughness and the cyclicfatigue resistance of epoxy-polymer blends. *Polymer* **2014**, *55*, 6325–6334. [CrossRef]
15. Schueler, R.; Petermann, J.; Schulte, K.; Wentzel, H.P. Agglomeration and electrical percolation behavior of carbon black dispersed in epoxy resin. *J. Appl. Polym. Sci.* **1997**, *63*, 1741–1746. [CrossRef]
16. Lan, T.; Pinnavaia, T.J. Clay-reinforced epoxy nanocomposites. *Chem. Mater.* **1994**, *6*, 2216–2219. [CrossRef]
17. Johnsen, B.B.; Kinloch, A.J.; Mohammed, R.D.; Taylor, A.C.; Sprenger, S. Toughening mechanisms of nanoparticle-modified epoxy polymers. *Polymer* **2007**, *48*, 530–541. [CrossRef]
18. Ajayan, P.M.; Schadler, L.S.; Giannaris, C.; Rubio, A. Single-walled carbon nanotubepolymer composites: Strength and weakness. *Adv. Mater.* **2000**, *12*, 750–753. [CrossRef]
19. Shen, M.Y.; Chang, T.Y.; Hsieh, T.H.; Li, Y.L.; Chiang, C.L.; Yang, H.; Yip, M.C. Mechanical Properties and Tensile Fatigue of Graphene Nanoplatelets Reinforced Polymer Nanocomposites. *J. Nanomater.* **2013**, 565401. [CrossRef]
20. Wang, P.N.; Hsieh, T.H.; Chiang, C.-L.; Shen, M.Y. Synergetic effects of mechanical properties on Graphene nanoplatelet and Multiwalled Carbon Nanotube Hybrids Reinforced Epoxy/Carbon Fiber Composites. *J. Nanomater.* **2015**, 838032. [CrossRef]
21. Hsieh, T.H.; Huang, Y.S.; Shen, M.Y. Carbon Nanotube Size Effect on the Mechanical Properties and Toughness of Nanocomposites. *Polym. Compos.* **2017**, *39*, E1072–E1086. [CrossRef]
22. Hsieh, T.H.; Chen, W.J.; Chiang, C.-L.; Shen, M.Y. Environmental aging effect on interlaminar properties of graphene nanoplatelets reinforced epoxy/carbon fiber composite laminates. *J. Reinf. Plast. Comp.* **2018**, *37*, 1177–1190. [CrossRef]
23. Kuan, C.F.; Chiang, C.-L.; Lin, S.H.; Huang, W.G.; Hsieh, W.Y.; Shen, M.Y. Characterization and Properties of Graphene Nanoplatelets/XNBR Nanocomposites. *Polym. Polym. Compos.* **2018**, *26*, 59–67. [CrossRef]
24. Zhang, X.C.; Peng, H.X.; Limmack, A.P.; Scarpa, F. Viscoelastic damping behaviour of cup stacked carbon nanotube modified epoxy nanocomposites with tailored interfacial condition and re-agglomeration. *Compos. Sci. Technol.* **2014**, *105*, 66–72. [CrossRef]
25. Wang, Y.; Zhang, J.; Shen, X.; Shi, C.; Wu, J.; Sun, L. Dispersion investigation of TiO2 nanoparticles coated by pulsed RF plasma polymer. *Mater. Chem. Phys.* **2006**, *98*, 217–224. [CrossRef]
26. Schilde, C.; Nolte, H.; Arlt, C.; Kwade, A. Effect of fluid–particle-interactions on dispersing nano-particles in epoxy resins using stirred-media-mills and three-roll-mills. *Compos. Sci. Technol.* **2010**, *70*, 657–663. [CrossRef]
27. Suihkonen, R.; Nevalainen, K.; Orell, O.; Honkanen, M.; Tang, L.; Zhang, H.; Zhang, Z.; Vuorinen, J. Performance of epoxy filled with nano- and micro-sized Magnesium hydroxide. *J. Mater. Sci.* **2012**, *47*, 1480–1488. [CrossRef]
28. Yasmin, A.; Luo, J.J.; Abot, J.L.; Daniel, I.M. Processing of expanded graphite reinforced polymer nanocomposites. *Compos. Sci. Technol.* **2006**, *66*, 1182–1189. [CrossRef]
29. Prolongo, S.G.; Moriche, R.; Jiménez-Suárez, A.; Sánchez, M.; Ureña, A. Advantages and disadvantages of the addition of graphene nanoplatelets to epoxy resins. *Eur. Polym. J.* **2014**, *61*, 206–214. [CrossRef]
30. Ahmadi-Moghadam, B.; Taheri, F. Fracture and toughening mechanisms of GNP-based nanocomposites in modes I and II fracture. *Eng. Fract. Mech.* **2014**, *131*, 329–339. [CrossRef]
31. Chandrasekaran, S.; Sato, N.; Tolle, F.; Mulhaupt, R.; Fiedler, B.; Schulte, K. Fracture toughness and failure mechanism of graphene based epoxy composites. *Compos. Sci. Technol.* **2014**, *97*, 90–99. [CrossRef]

32. Mannov, E.; Schmutzler, H.; Chandrasekaran, S.; Viets, C.; Buschhorn, S.; Tolle, F.; Mulhaupt, R.; Schulte, K. Improvement of compressive strength after impact in fibre reinforced polymer composites by matrix modification with thermally reduced graphene oxide. *Compos. Sci. Technol.* **2013**, *87*, 36–41. [CrossRef]
33. Prolongo, S.G.; Jimenez-Suarez, S.; Moriche, R.; Urena, A. In situ processing of epoxy composites reinforced with graphene nanoplatelets. *Compos. Sci. Technol.* **2013**, *86*, 185–191. [CrossRef]
34. Kostagiannakopoulou, C.; Loutas, T.H.; Sotiriadis, G.; Markou, A.; Kostopoulos, V. On the interlaminar fracture toughness of carbon fiber composites enhanced with graphene nano-species. *Compos. Sci. Technol.* **2015**, *118*, 217–225. [CrossRef]
35. Chatterjee, S.; Nafezarefi, F.; Tai, N.H.; Schlagenhauf, L.; Nüesch, F.A.; Chu, B.T.T. Size and synergy effects of nanofiller hybrids including graphene nanoplatelets and carbon nanotubes in mechanical properties of epoxy composites. *Carbon* **2012**, *50*, 380–5386. [CrossRef]
36. Bridgwater, J. Mixing of powders and granular materials by mechanical means—A perspective. *Particuology* **2012**, *10*, 397–427. [CrossRef]
37. Park, Y.M.; Lee, D.H.; Hwang, W.R.; Lee, S.B.; Jung, S.I. Hydrodynamics of CNT dispersion in high shear dispersion mixers. *Korea-Aust. Rheol. J.* **2014**, *26*, 347–353. [CrossRef]
38. Takatsuka, T.; Endo, T.; Jianguo, Y.; Yuminoki, K.; Hashimoto, N. Nanosizing of poorly water soluble compounds using rotation/revolution mixer. *Chem. Pharm. Bull.* **2009**, *57*, 1061–1067. [CrossRef]
39. Tsuchiya, K.; Sakai, A.; Nagaoka, T.; Uchida, K.; Furukawa, T.; Yajima, H. High electrical performance of carbon nanotubes/rubber composites with low percolation threshold prepared with a rotation-revolution mixing technique. *Compos. Sci. Technol.* **2011**, *71*, 1098–1104. [CrossRef]
40. Mizuuchi, K.; Inoue, K.; Agari, Y.; Sugioka, M.; Tanaka, M.; Takeuchi, T.; Kawahara, M.; Makino, Y.; Ito, M. Processing of diamond-particle-dispersed silver-matrix composites in solid-liquid co-existent state by SPS and their thermal conductivity. *Compos. B Eng.* **2012**, *43*, 1445–1452. [CrossRef]
41. Galpaya, D.; Wang, M.; George, G.; Motta, N.; Waclawik, E.; Yan, C. Preparation of graphene oxide/epoxy nanocomposites with significantly improved mechanical properties. *J. Appl. Phys.* **2014**, *116*, 053518. [CrossRef]
42. Cha, J.; Jun, G.H.; Park, J.K.; Kim, J.C.; Ryu, H.J.; Hong, S.H. Improvement of modulus, strength and fracture toughness of CNT/Epoxy nanocomposites through the functionalization of carbon nanotubes. *Compos. B Eng.* **2017**, *129*, 169–179. [CrossRef]
43. Cha, J.; Kim, J.; Ryu, S.; Hong, S.H. Comparison to mechanical properties of epoxy nanocomposites reinforced by functionalized carbon nanotubes and graphene nanoplatelets. *Compos. B Eng.* **2019**, *162*, 283–288. [CrossRef]
44. Chiang, C.-L.; Yeh, Y.L.; Hsu, C.W.; Shen, M.Y. Dispersion Technology on Mechanical Properties of Carbon nanomaterials Reinforced Epoxy Nanocomposites. *Mod. Phys. Lett. B* **2020**, *34*, 2040019. [CrossRef]
45. Kim, S.H.; Kim, S.J.; Park, S.J.; Mun, J.H.; Kang, T.G.; Park, J.M. Rheological behavior of magnetic powder mixtures for magnetic PIM. *Korea-Aust. Rheol. J.* **2012**, *24*, 121–127. [CrossRef]
46. Park, J.M.; Park, S.J. Rheology and processing of suspensions with fiber, disk and magnetic particles. *Korea-Aust. Rheol. J.* **2011**, *23*, 219–226. [CrossRef]
47. Son, K.J. A discrete element model for the influence of surfactants on sedimentation characteristics of magnetorheological fluids. *Korea-Aust. Rheol. J.* **2018**, *30*, 29–39. [CrossRef]
48. Graphene-Wikipedia. Available online: https://en.wikipedia.org/wiki/Graphene (accessed on 20 January 2021).
49. Heidenreich, R.D.; Hess, W.M.; Ban, L.L. A test object and criteria for high resolution electron microscopy. *J. Appl. Crystallogr.* **1968**, *1*, 1–19. [CrossRef]
50. Gojny, F.H.; Wichmann, M.H.G.; Köpke, U.; Fiedler, B.; Schulte, K. Carbon nanotube-reinforced epoxy-composites: Enhanced stiffness and fracture toughness at low nanotube content. *Compos. Sci. Technol.* **2004**, *64*, 2363–2371. [CrossRef]
51. Chen, X.; Dobson, J.F.; Raston, C.L. Vortex fluidic exfoliation of graphite and boron nitride. *Chem. Commun.* **2012**, *48*, 3703–3705. [CrossRef] [PubMed]

sustainability MDPI

Article

Flame Retardance and Char Analysis of an Eco-Friendly Polyurethane Hyperbranched Hybrid Using the Sol–Gel Method

Ming-Yuan Shen [1], Chen-Feng Kuan [2], Hsu-Chiang Kuan [2], Cing-Yu Ke [3] and Chin-Lung Chiang [3,*]

[1] Department of Mechanical Engineering, National Chin-Yi University of Technology, Taichung City 41170, Taiwan; hbj678@gmail.com
[2] General Education Center, Far East University, Tainan City 74448, Taiwan; cfkuan@mail.feu.edu.tw (C.-F.K.); hckuan@mail.feu.edu.tw (H.-C.K.)
[3] Green Flame Retardant Material Research Laboratory, Department of Safety, Health and Environmental Engineering, Hung-Kuang University, Taichung City 43302, Taiwan; h22432003@gmail.com
* Correspondence: dragon@sunrise.hk.edu.tw; Tel.: +886-4-26318652 (ext. 4008)

Abstract: This study used the sol–gel method to synthesize a non-halogenated, hyperbranched flame retardant containing nitrogen, phosphorus, and silicon (HBNPSi), which was then added to a polyurethane (PU) matrix to form an organic–inorganic hybrid material. Using ^{29}Si nuclear magnetic resonance, energy-dispersive X-ray spectroscopy of P- and Si-mapping, scanning electron microscopy, and X-ray photoelectron spectroscopy, this study determined the organic and inorganic dispersity, morphology, and flame retardance mechanism of the hybrid material. The condensation density of the hybrid material PU/HBNPSi was found to be 74.4%. High condensation density indicates a dense network structure of the material. The P- and Si-mapping showed that adding inorganic additives in quantities of either 20% or 40% results in homogeneous dispersion of the inorganic fillers in the polymer matrix without agglomeration, indicating that the organic and inorganic phases had excellent compatibility. In the burning test, adding HBNPSi to PU made the material pass the UL-94 test at the V2 level, unlike the pristine PU, which did not meet the standard. The results demonstrate that after non-halogenated flame retardant was added to PU, the material's flammability and dripping were lower, thereby proving that flame retardants containing elements such as nitrogen, phosphorus, and silicon exert an excellent flame-retardant synergistic effect.

Keywords: flame retardant; organic–inorganic hybrid; polyurethane; sol–gel method; synergistic effect

Citation: Shen, M.-Y.; Kuan, C.-F.; Kuan, H.-C.; Ke, C.-Y.; Chiang, C.-L. Flame Retardance and Char Analysis of an Eco-Friendly Polyurethane Hyperbranched Hybrid Using the Sol–Gel Method. *Sustainability* **2021**, *13*, 486. https://doi.org/10.3390/su13020486

Received: 18 December 2020
Accepted: 30 December 2020
Published: 6 January 2021

1. Introduction

Polyurethane (PU) can appear in many forms, such as foam, adhesive, paint, and elastomer. Because of its favorable mechanical properties and resistance to chemicals and wear, PU is widely used in the automobile, textile, sports equipment, shoe sole, and paint industries, among others [1–3]. PU is the most common commercial polymer material in daily life. However, it has some weaknesses, such as low thermal stability and high flammability. For example, when PU is heated to 80–90 °C, its mechanical properties rapidly deteriorate. When the PU matrix is heated to over 200 °C, severe pyrolysis occurs and it becomes flammable, increasing the risk of a fire hazard. Therefore, improving the flame retardance of PU is a critical task in the development of polymer materials [1,4,5].

Compared with halogenated flame retardants, non-halogenated flame retardants have attracted great attention because during their combustion a lower amount of hazardous gases or smoke is produced, which is safer for the people in the fire and will not destroy the ozone layer in the atmosphere. These are called eco-friendly flame retardants. Flame retardants can play a role in the gas phase or condensed phase. Halogenated flame retardants are typically in the gas phase, whereas flame retardants containing phosphorus and nitrogen can be in either phase [4,6,7]. Flame retardants containing siloxane have a main chain of Si–O. The energy of the Si–O bond is high and the Si–O chain is resilient when forming SiO_2; as a result, these flame retardants exhibited outstanding thermal stability under high temperatures [8]. Rao et al. [9] reported an organic–inorganic phosphorus–nitrogen–silicon flame retardant that was synthesized by the Kabachnik–Fields reaction and the sol–gel method, and then it was used as a reactive flame retardant to prepare flame-retardant and smoke-suppressant epoxy resins (EPs). The results certified the charring effect of the phosphaphenanthrene group and the enhancing effect of the silicon group. Subsequently, the flame inhibition effect and lesser combustible gases release enhanced the flame-retardant properties of the EP.

In the last few decades, the hyperbranched hybrid has gained more attention due to its special structure. Many hyperbranched materials have been synthesized and used in many fields, including flame retardants. Yang et al. [10] reported a novel hyperbranched phosphorus/nitrogen-containing flame retardant (HPNFR) that was facilely synthesized via the transesterification reaction of dimethyl methylphosphonate and tris (2-hydroxyethyl) isocyanurate. The sample with 4 wt% HPNFR can achieve a V-0 rating in the UL-94 test and possesses a limiting oxygen index (LOI) value as high as 34.5%. HPNFR would not significantly damage the transparence of EP thermosets; consequently, it reserved its application value in some special fields. Hu et al. [11] showed a phosphorus/nitrogen-containing hyperbranched polymer (PN-HBP) that was synthesized via the esterification reaction of 2-carboxyethyl (phenyl) phosphinic acid (CEPPA) and tris (2-hydrooxyethyl) isocyanurate (THEIC). A higher LOI and a V-0 rating in the UL-94 vertical burning test were realized, which indicated an apparent synergistic effect. The peak heat release rates (PHRRs) of composites were reduced significantly compared with that of PU.

The PU used in this study has a wide range of applications. However, it has fatal disadvantages such as flammability and dripping. This study employed the sol–gel method to prepare a reactive flame retardant containing nitrogen, phosphorus, and silicon and classified the flame retardance of the organic–inorganic polyurethane hybrid.

2. Experimental

2.1. Materials

Isophorone diisocyanate (IPDI) and dibutyltin dilaurate (DBTDL) were purchased from Alfa Aesar Co., Shore Road, Heysham LA3 2XY, UK. ARCOL POLYOL 1021 (Polyol) was purchased from Bayer Material Science Taiwan Limited Taipei, Taiwan. The purchase of 1,4-Butanediol (1,4-BD) was from Acros Chemical Co., Morristown, NJ, USA. The purchase of 3-Isocyanatopropyltriethoxysilane (IPTS) was from GELEST, Inc., 11 steel Rd. EAST, Morrisville, PA, USA. Triglycidyl isocyanurate (TGIC) was purchased from TCI, Tokyo, Japan. The purchase of 10-(2,5-Dihydroxyphenyl)-10H-9-oxa-10-phosphaphenanthbrene-10-oxide (DOPO-BQ) was from Sigma-Aldrich Co. Ltd., Taipei, Taiwan. Anhydrous stabilized tetrahydrofuran (THF) was obtained from Lancaster Co., Morecambe, Lancashire, UK. Hydrogen chloride (HCl) was purchased from ECHO Chemical Co., Ltd., Toufen, Taiwan.

2.2. Preparation of DOPOBQ-IPTS-TGIC

First, DOPO-BQ (7.58 g) and IPTS (5.78 g) were added into a 100 mL serum bottle, and then 80 mL of THF solvent and 0.2 g of metal catalyst DBTDL were added. The aforementioned reactants were stirred at 60 °C and the substances were allowed to react for 2 h. Subsequently, we added 0.72 mL of deionized (DI) water into THF, instilled HCl, and adjusted the pH value to 4; the product of the sol–gel reaction was hyperbranched phosphorous–silicon (HBPSi). Additionally, TGIC (6.95 g) was dissolved in THF and a sol–gel reaction was conducted for 2 h on HBPSi. After the temperature reached 60 °C, TGIC was slowly added into HBPSi and allowed to react at 60 °C for 2 h. This formed hyperbranched nitrogen-phosphorous-silicon (HBNPSi) flame retardant that was dissolved in THF and then the addition reaction took place with prepolymer to form nanocomposites. This reaction is presented in Scheme 1.

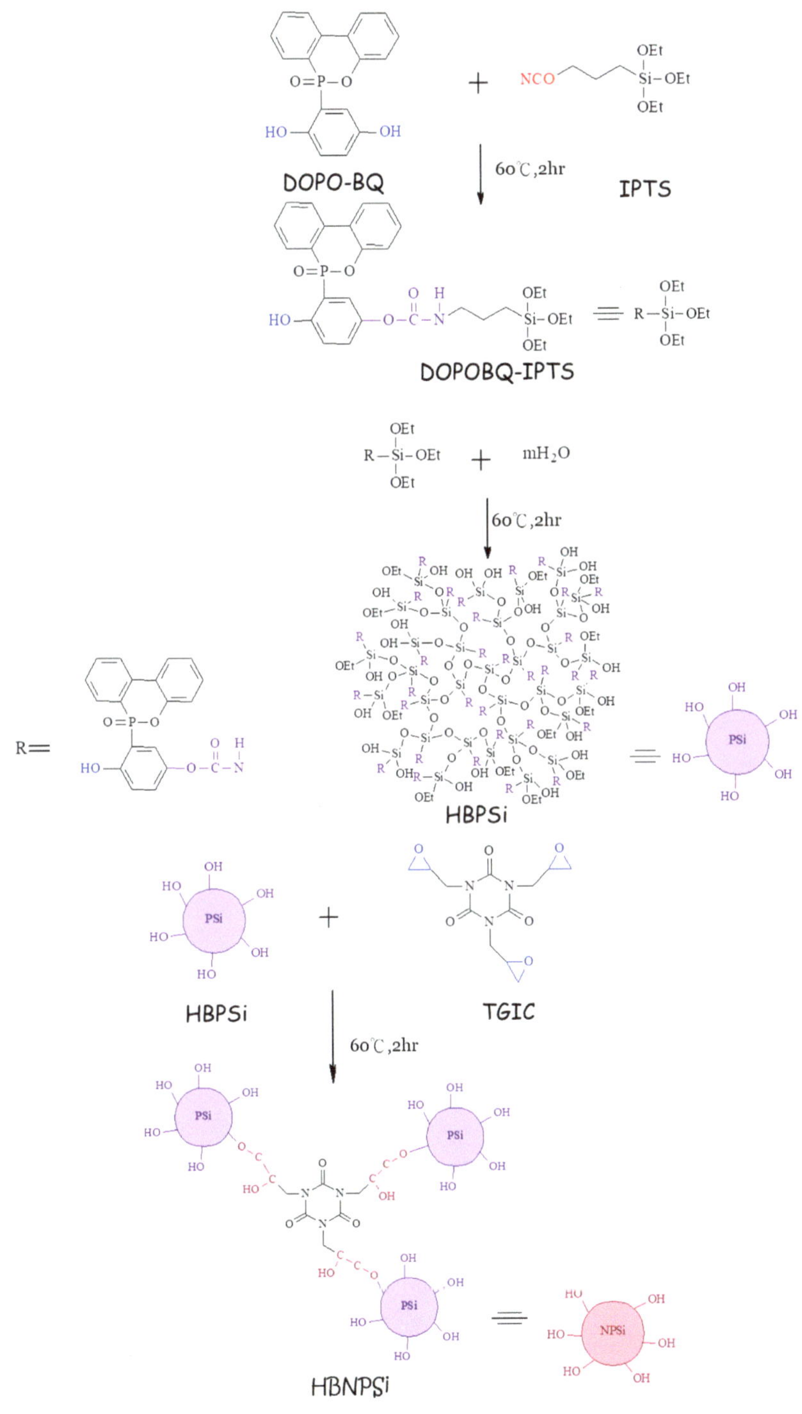

Scheme 1. The reaction of hyperbranched nitrogen–phosphorous–silicon (HBNPSi).

2.3. Preparation of PU/HBNPSi Hybrid

IPDI (9.88 g) and Polyol (20 g) were added into a four-necked bottle in a nitrogen environment; the mixture was stirred by a magnetic stirrer (PC 410 D, Corning Incorporated, Corning, NY, USA), at 60 °C. Subsequently, DBTDL catalyst (0.3 g) was added into the four-necked bottle for a simultaneous reaction to form a prepolymer. Following 2 h of reaction, HBNPSi was slowly dripped into the four-necked bottle. After 2 h of reaction, 1,4-BD (0.8 g) was also allowed to react for 2 h and we observed whether the viscosity increased. After the viscosity increased for any sample, that sample was poured into a polytetrafluoroethylene mold and placed into a vacuum oven (DOV-40, Dengyng Co., New Taipei City, Taiwan) for 24 h of de-aeration. After the temperature reached 60 °C, it was placed in another circulating oven (OV-80, Firstek Scientific Co., Hsin Chuang City, Taiwan) for 24 h. When the temperature reached 60 °C, the finished product was withdrawn and stored at room temperature for cooldown, thereby completing the preparation of PU/HBNPSi. Its reaction formula is presented in Scheme 2. The schematic diagram for the experimental part is shown in Scheme 3.

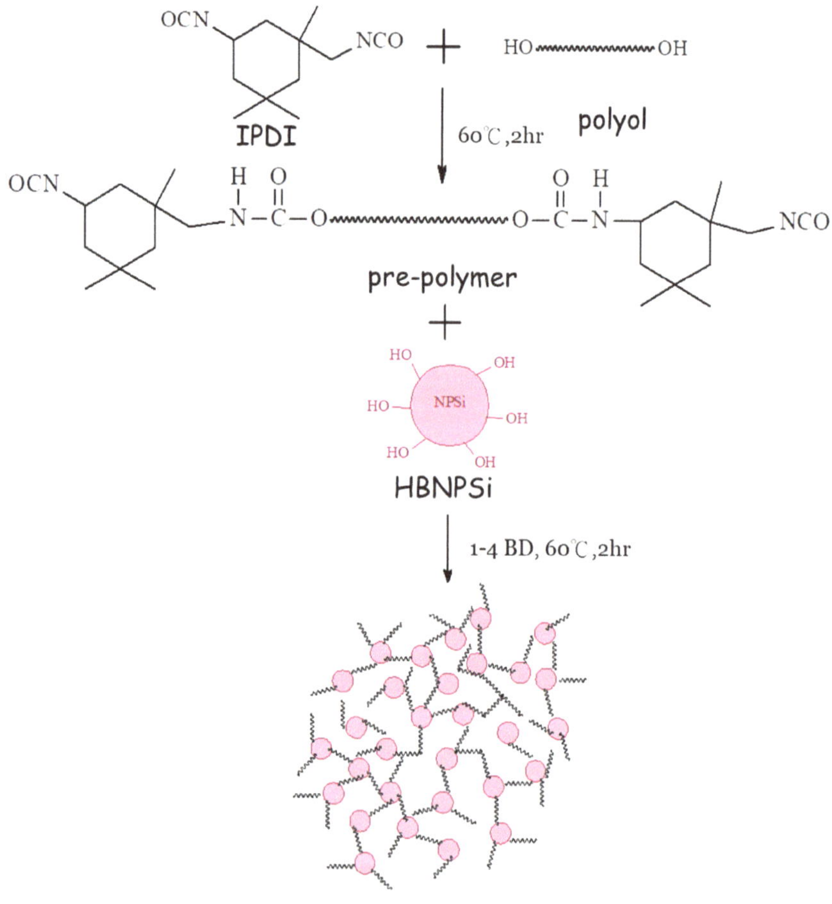

Scheme 2. The reaction of polyurethane (PU)/HBNPSi.

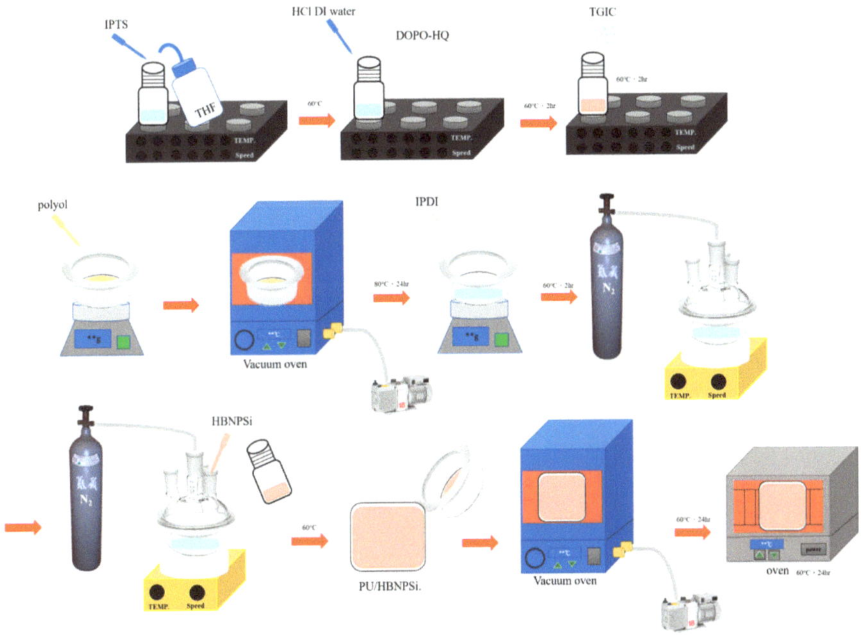

Scheme 3. The schematic diagram for the experimental part.

2.4. Measurements

To observe the condensation density, ^{29}Si nuclear magnetic resonance (NMR) was used and performed by a Bruker DSX-400WB, Bremen, Germany. The samples were treated at 180 °C for 2 h and then ground into a fine powder. The vertical burning test was done inside a fume hood. The UL-94 standard was then used to evaluate the flame retardance of the hybrid material. Samples were held vertically with tongs at one end and burned from the free end. Three samples were prepared for every test. Samples were exposed to an ignition source for 10 s. Then, they were allowed to burn above cotton wool until both the sample and cotton wool were extinguished. Observable parameters were recorded to assess fire retardance. The UL-94 test classifies the materials as V-0, V-1, and V-2 according to the period needed before self-extinction and the occurrence of flaming dripping after removing the ignition source. V-0 is the most ambitious and desired classification. X-ray photoelectron spectroscopy (XPS) was adopted to determine the char composition to observe the mechanism of burning. The process for using the high-resolution X-ray photoelectron spectrometer (HR-XPS, ULVAC-PHI, Inc., Kanagawa-ken, Japan) is as follows: the sample is crushed into powder and then adhered to the aluminum sheet with small round holes. This sheet is mainly used for detecting the sample's surface as well as its elemental composition and distribution in vertical directions, in addition to implementing analysis on the links of elemental substances. The morphology of the fractured surface of the composites was studied under a scanning electron microscope (SEM) (JEOL JSM 840A, Tokyo, Japan). Energy-dispersive X-ray spectroscopy (EDX) was used to observe the distributions of Si and P atoms in the hybrid, which were obtained from SEM EDX mapping (JEOL JSM 840A, Tokyo, Japan).

3. Results and Discussion

3.1. ^{29}Si NMR

Solid-state ^{29}Si NMR spectrometry was used to determine the structure of the PU/HBNPSi hybrid material, and the DOPO-BQ-IPTS reaction process and its degree of hydrolysis condensation was obtained using the sol–gel method. Because one end of IPTS was terminated with three tri-alkoxy groups (T), the end of the prepolymer had a T-shaped structure. The other end was terminated with an organic chain—an –NCO functional group—that reacted with the –OH functional group in DOPO-BQ. The T-end further hydrolyzed and condensed to form a network structure of Si–O–Si, which was more stable. According to the degree of hydrolysis and condensation, the single-, double-, and triple-replacement absorption peaks located at −45 to −48 ppm, −56 to −62 ppm, and −66 to −69 ppm indicate the T^1 [12,13], T^2 [12,14], and T^3 [12,13] structures, respectively.

Figure 1 presents the solid-state ^{29}Si NMR spectra of the PU/HBNPSi hybrid material. The hybrid material had the T-structure, primarily T^3. Using peak separation treatments, the T^1, T^2, and T^3 areas could be identified. Then, the following equation was used to calculate the condensation density (Dc(%)) [13].

$$\mathrm{Dc}(\%) = \frac{1 \times \left(\%\mathrm{areaT}^1\right) + 2 \times \left(\%\mathrm{areaT}^2\right) + 3 \times \left(\%\mathrm{areaT}^3\right)}{3}$$

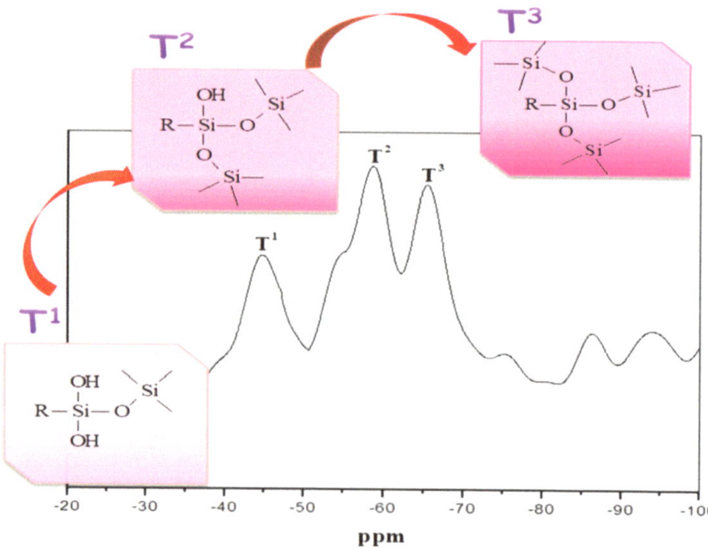

Figure 1. Solid-state ^{29}Si nuclear magnetic resonance (NMR) spectra of PU/HBNPSi.

The Dc(%) of the hybrid material PU/HBNPSi was calculated to be 74.4%. Higher Dc(%) indicates that a material has a denser network structure. The Si–O–Si bonded into an excellent network structure, and the Si–O bond had high bond energy. The hybrid material thus had high thermal stability and flame retardance. The results are summarized in Table 1.

Table 1. Distribution of area proportion.

Sample NO.	Area (%)		
	T^1	T^2	T^3
PU/HBNPSi 40%	24.3	44.3	36.8

3.2. P- and Si-Mapping of EDX

This study determined the compatibility of pristine PU and HBNPSi. Two phases were obtained and demonstrated differences under different interaction forces. The P- and

Si-mapping of energy-dispersive X-ray spectroscopy was utilized to determine whether the dispersibility and homogeneity were favorable or whether agglomeration occurred when the inorganic phase was mixed with the organic phase. Favorable dispersibility indicates that the two phases have excellent compatibility, which is crucial in a hybrid material.

Figure 2 displays P- and Si-mapping of 20% and 40% PU/HBNPSi. The images show the dispersion of the inorganic elements P and Si in their organic phases. Each bright dot represents P or Si. The Figure reveals that the inorganic substances in both the 20% and 40% PU/HBNPSi were dispersed favorably in the matrix and that the material was homogeneous. No agglomeration was observed, which indicated that the compatibility between the organic and inorganic phases was excellent. The amount of P and Si in the material increased as the HBNPSi content was increased.

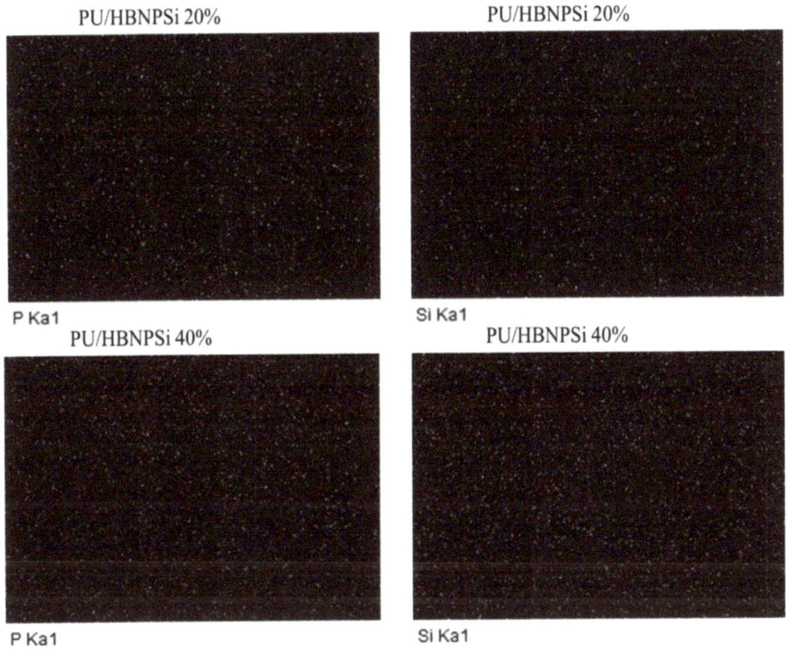

Figure 2. Mapping photograph of the P- and Si-containing hybrid PU/HBNPSi.

3.3. Flame Retardance Analysis

In the UL-94 test, a standardized ignition specimen is combusted and the total burning duration (the sum of two burning durations) must meet the standard; additionally, during the burning process, dripping, which could ignite the cotton beneath the specimen, must not occur. The flame retardance of polymeric material is classified into the V-0, V-1, and V-2 levels. Figure 3 and Table 2 reveal that the pristine PU failed and did not achieve any level. Dripping was observed, and the drips ignited the cotton below. As the concentration of HBNPSi in the material was increased to 40%, the two flame burning durations of the PU/HBNPSi hybrid material were 0.7 and 16.1 s, respectively. The sum of these two burning durations was 16.8 s. However, because dripping occurred, the material was graded as V-2. HBNPSi had excellent flame retardance because during pyrolysis the N and P in the structure became gaseous, capturing free radicals and playing a role in the condensed phase, thereby catalyzing the formation of char. Additionally, Si migrated across the surface [14,15] and formed a dense SiO_2 structure that protected the interior of the materials. The benzene ring of DOPO-BQ provided the source for char formation, and the char layer resisted burning under high temperature.

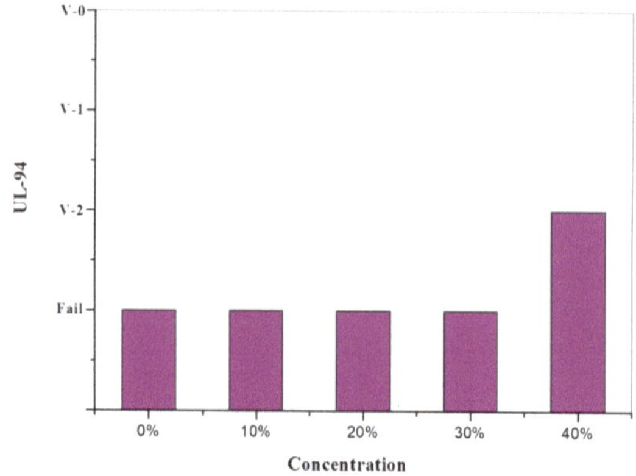

Figure 3. Effect of various HBNPSi concentrations on the UL-94 of the PU/HBNPSi hybrid.

Table 2. The flame retardance of the PU/HBNPSi hybrid by UL-94 values.

Sample	UL-94		
	$t_1 + t_2$ (s)	Ranking	Dripping
Pristine PU	>30	Fail	Yes
PU/HBNPSi 10%	>30	Fail	Yes
PU/HBNPSi 20%	>30	Fail	Yes
PU/HBNPSi 30%	>30	Fail	Yes
PU/HBNPSi 40%	16.8	V2	Yes

3.4. Morphology Analysis

SEM was used to observe the morphology of the hybrid material and determine its surface microtopography, particle size, and surface nucleation. Figure 4a shows that the surface of the pristine PU was smooth and homogeneous before burning, without phase separation. Figure 4b displays the surface of the burned pristine PU. Because pristine PU is highly flammable and molten after burning, some wave patterns can be observed in the SEM image. Figure 4c shows the surface of the 20% PU/HBNPSi before burning; it was generally smooth and without phase separation, despite the addition of HBNPSi. Figure 4d displays the surface of 20% PU/HBNPSi after burning. Some Si particles can be seen because Si migrated across the surface during burning. However, because insufficient char was formed, the heat blocking effect was weak, calling to mind UL-94. Therefore, 20% PU/HBNPSi was flammable. Figure 4e shows the surface of 40% PU/HBNPSi before burning and that the surface had discrete HBNPSi particles. Although the additional loading was 40%, no phase separation was observed. Figure 4f shows the surface of 40% PU/HBNPSi after burning. A dense char layer covered this surface because P was dehydrated and catalyzed into char and Si migrated across the surface while burning to form a dense char layer through the condensed phase. Both P and Si covered the material surface [13–15], which blocked the transmission of gas and flame and increased thermal stability.

3.5. XPS Char Analysis

XPS was used to examine the changes in chemical bonds in the PU/HBNPSi hybrid material at room temperature and after being burned at 800 °C in a high-temperature furnace. Additionally, after flame retardant was added to the PU the functional group changes, before and after the hybrid was burned, were observed. The peak separation method was used to calculate the material's anti-oxidation characteristics. The results are presented in Figures 5–10 and Tables 3 and 4.

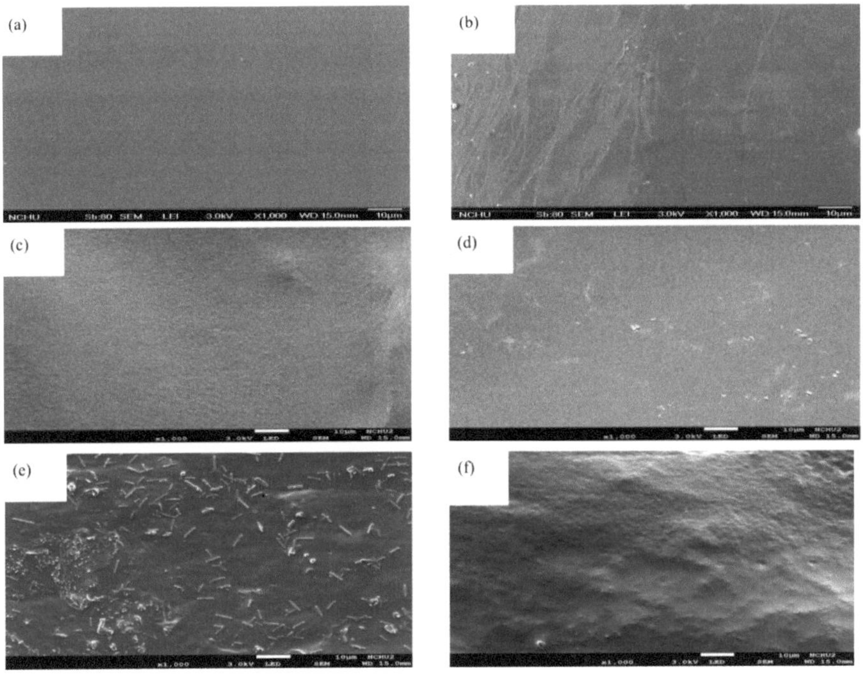

Figure 4. SEM micrographs of composites (**a**) pristine PU (before burning) (×1 K), (**b**) pristine PU (after burning) (×1 K), (**c**) PU/HBNPSi 20% (before burning) (×1 K), (**d**) PU/HBNPSi 20% (after burning) (×1 K), (**e**) PU/HBNPSi 40% (before burning) (×1 K), (**f**) PU/HBNPSi 40% (after burning) (×1 K).

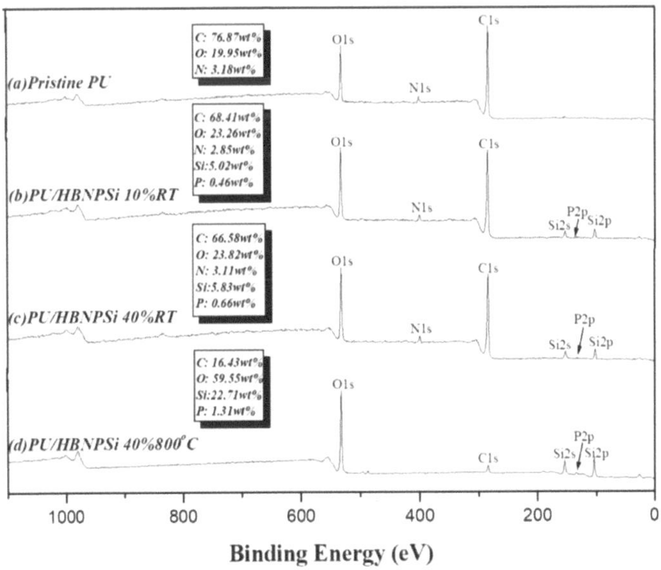

Figure 5. XPS survey spectra of (**a**) pristine PU, (**b**) PU/HBNPSi 10% at RT, (**c**) PU/HBNPSi 40% at RT, (**d**) PU/HBNPSi 40% at 800 °C.

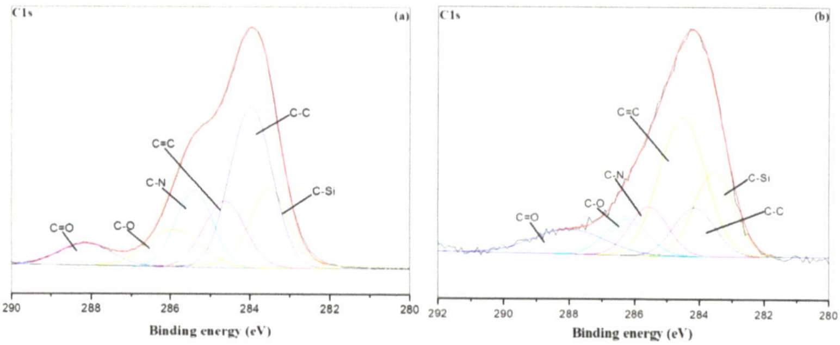

Figure 6. C1s spectra of PU/HBNPSi 10%: (**a**) RT; (**b**) under an air atmosphere at 800 °C for 30 min.

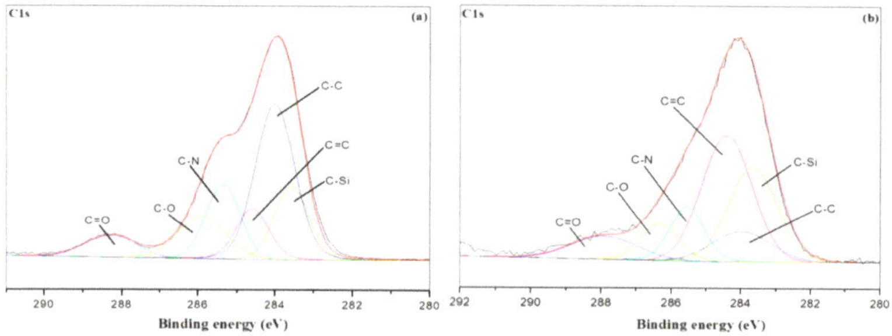

Figure 7. C1s spectra of PU/HBNPSi 40%: (**a**) RT; (**b**) under an air atmosphere at 800 °C for 30 min.

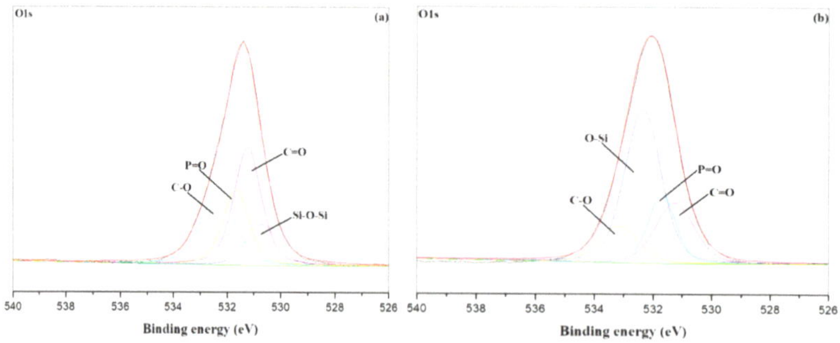

Figure 8. O1s spectra of PU/HBNPSi 40%: (**a**) RT; (**b**) under an air atmosphere at 800 °C for 30 min.

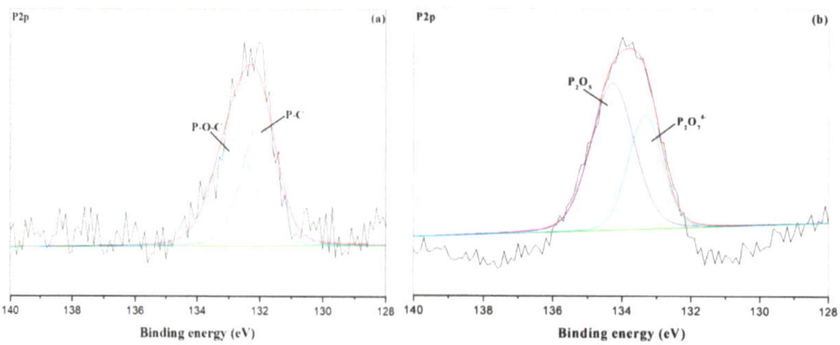

Figure 9. P2p spectra of PU/HBNPSi 40%: (**a**) RT; (**b**) under an air atmosphere at 800 °C for 30 min.

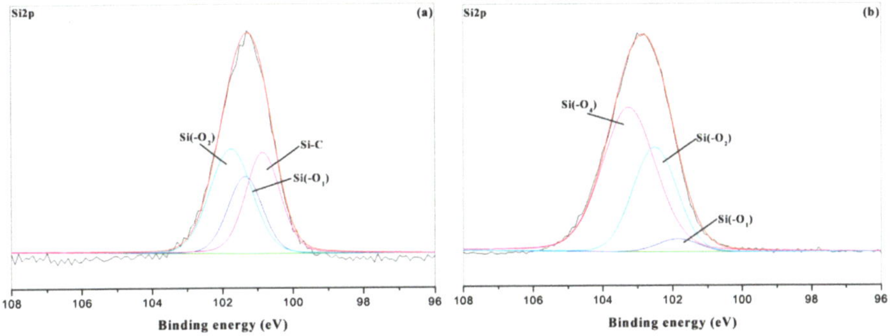

Figure 10. Si2p spectra of PU/HBNPSi 40%: (**a**) RT; (**b**) under an air atmosphere at 800 °C for 30 min.

Table 3. Binding energy (eV) and relative peak intensities (%) of the various components of C1s peak-fitted signals.

Sample NO.	C1s					
	C–C/C–H	C-Si	C=C	C–N	C–O	C=O
PU/HBNPSi 10%—RT	0.31	0.16	0.11	0.12	0.15	0.08
PU/HBNPSi 10%—800 °C	0.11	0.20	0.37	0.11	0.09	0.10
PU/HBNPSi 40%—RT	0.38	0.15	0.10	0.15	0.13	0.08
PU/HBNPSi 40%—800 °C	0.14	0.38	0.48	0.12	0.05	0.07

Table 4. The values of Cox/Ca of composites at room temperature (RT) and 800 °C.

Sample NO.	Temperature	
	RT	800 °C
PU/HBNPSi 10%	0.33	0.24
PU/HBNPSi 40%	0.27	0.11

Figure 5 displays XPS rough scans illustrating the elemental composition of the pristine PU, 10% PH/HBNPSi at room temperature, 40% PH/HBNPSi at room temperature, and 40% PU/HBNPSi after being burned at 800 °C. The scan in Figure 5a indicates three elements: C, O, and N. For the hybrid material containing 10% flame retardant (Figure 5b), P and Si were observed in addition to C, O, and N. The XPS rough scan, obtained after DOPO-BQ and IPTS were reacted, verifies that the hybrid material contained P and Si. Figure 5c presents the rough scan for the hybrid material containing 40% of HBNPSi. Because the additive content was higher than for the 10% hybrid material, the P and Si content were higher. Figure 5d shows that the hybrid material was 40% HBNPSi after being burned. Burning at a high temperature led to a slight increase in the amount of P and Si in the hybrid material because during the burning process P was dehydrated and formed char and Si migrated across the surface [14,15]. Together, these elements formed a layer of char containing P and Si that protected the material. Nitrogen mainly exerts a flame-retardant effect in the gas phase, which can dilute the oxygen concentration and capture free radicals in the gas phase, so in the XPS spectrum, the nitrogen element disappears in Figure 5d.

Figures 6–10 present scans showing Cls, Ols, P2p, and Si2p bonds in the materials at room temperature and after being burned at 800 °C. Six functional groups can be observed in the Cls spectra: those at 284 eV (C–C and C–H) [16], 285.4 eV (C–N) [17], 288.5 eV (C=O) [18], 284.5 eV (C=C) [19], and 286 eV (C–O) [20]. Because the material additive contained silicide, the spectrum contains a peak at 282.9 eV (C–Si) [21]. After heat oxidation at 800 °C, the material was mostly graphitized. The C=C content was higher, as illustrated in Figures 6 and 7. The formation of graphitized carbon–carbon double bonds is very important for the improvement of thermal stability. It takes more than 3000 degrees to destroy its structure and form a high thermal stability protective layer on the surface of the material.

Figure 8 shows the Ols spectra, in which four functional groups can be observed, namely 531.2 eV (C=O), 532.2 eV (C–O), 531.5 eV (P=O), and 530.6 eV (Si–O) [22–27]. This indicates the change in bonding type after high-temperature heat oxidation. In Figure 8a, the Si–O–Si bonds are formed by the sol–gel reaction. In Figure 8b, it can be clearly seen that a large number of Si–O bonds are formed, mainly to form silicon dioxide, which is resistant to high temperatures. The heat-resistant layer will protect the polymer matrix on the surface of the material.

Figure 9 shows that the P2p spectrum has peaks at 131.5 eV (P–C) [28] and 133 eV (P–O–C) [27]. The corresponding bonds formed because of the addition of DOPO-BQ flame retardant. After high-temperature deep oxidation, the peaks were at 135 eV (P_2O_5) and 133.1 ± 0.3 eV ($P_2O_7^{4-}$) [29], corresponding to inorganic phosphides, forming char protective material. The two aforementioned inorganic phosphides were converted from the bonds before burning. P_2O_5 is a kind of glass, which is a non-combustible material. It will melt and cover the surface of the material to achieve insulation resistance and material transfer, which can improve the flame-retardant properties of the material.

The Si2p spectra presented in Figure 10a,b, obtained before and after burning, respectively, have considerable differences in bonding strength. Figure 10a reveals that the siloxane coupling agent caused the hydrolysis condensation reaction. Three bonds were identified: those corresponding to peaks at 100.7 eV (Si–C), 101.6 eV (Si(–O_1)), and 102.2 eV (Si(–O_2)). After burning and oxidation, the Si–C peak has disappeared whereas peaks have appeared at 101.6 eV (Si(–O_1)), 102.2 eV (Si(–O_2)), and 103.5 eV (Si(–O_4)) [30] (Figure 10b). Because silicide was converted into SiO_2, the peak intensity of Si(–O_4) increased. This result indicates that after burning, the hybrid material PU/HBNPSi had char layers that covered and protected the matrix. Silica is an inorganic glass that will not burn and will melt at high temperatures, covering the surface of the material to resist high-temperature attacks.

We further analyzed the differences between PU/HBNPSi 10% and PU/HBNPSi 40% at room temperature and 800 °C for 30 min under an air atmosphere. After calculating the area ratio of each bond species individually using the Cls spectrum from Figure 6 to Figure 7, the Cox (oxidized carbons)/Ca (aliphatic, aromatic carbons) ratio was obtained to show the antioxidant effect of the material [31]. The results are shown in Tables 3 and 4. The ratio of Cox/Ca at room temperature of the mixed material with 10% and 40% concentration are 0.33 and 0.27, respectively. After high-temperature combustion, the ratio decreases to 0.24 and 0.11, respectively. In summary, the higher the concentration, the better the oxidation resistance, thus improving the thermal stability of hybrid materials.

4. Conclusions

This study used ^{29}Si NMR to determine that the siloxane condensation degree of the PU/HBNPSi was 74.4%. High condensation density indicates superior material properties. P- and Si-mapping revealed that the additive HBNPSi was well dispersed and homogeneously distributed in the polymer matrix. When the HBNPSi concentration was increased, aggregation did not occur, showing that the materials and additives were compatible. The UL-94 standard was used to determine the flame retardancy of the hybrid material. Pristine PU failed to meet the standard. The 40% PU/HBNPSi, however, achieved the V-2 level, proving that the flame retardant effectively improved the flame retardancy of the pristine PU. Using XPS, char analysis was conducted on the hybrid material PU/HBNPSi to determine the correlation between char formation and the material's thermal property. Cls, Ols, and Si2p XPS spectra revealed that the phosphide and siloxane formed by burning had the effect of protecting the material. A higher concentration of HBNPSi resulted in a stronger effect, achieving high thermal stability and an antioxidation effect. The organic–inorganic hybrid material prepared by the sol–gel method has good compatibility and flame resistance, and has great potential for future applications in coating applications.

Author Contributions: Conceptualization, C.-L.C.; Methodology & Software, M.-Y.S.; Validation, C.-F.K., H.-C.K., and C.-L.C.; Formal Analysis, C.-L.C.; Investigation, C.-Y.K.; Data Curation, Writing—Original Draft Preparation, Writing—Review & Editing, C.-L.C. All authors have read and agreed to the published version of the manuscript.

Funding: The authors would like to express their appreciation for the National Science Council of the Republic of China for the financial support of this study under grant MOST-108-2221-E-241-007.

Institutional Review Board Statement: Not applicable for studies not involving humans.

Informed Consent Statement: Not applicable for studies not involving humans or animals.

Data Availability Statement: Not applicable.

Conflicts of Interest: The authors declare no conflict of interest.

References

1. Kumar, R.; Yadav, R.; Kolhe, M.A.; Bhosale, R.S.; Narayan, R. 8-Hydroxypyrene-1,3,6-trisulfonic acid trisodium salt (HPTS) based high fluorescent, pH stimuli waterborne polyurethane coatings. *Polymer* **2018**, *136*, 157–165. [CrossRef]
2. Bedő, D.; Imre, B.; Domján, A.; Schöne, P.; Vancsoe, G.J.; Pukánszky, B. Coupling of poly(lactic acid) with a polyurethane elastomer by reactive processing. *Eur. Polym. J.* **2017**, *97*, 409–417. [CrossRef]
3. Zhu, S.W.; Shi, W.F. Flame retardant mechanism of hyperbranched polyurethane acrylates used for UV curable flame retardant coatings. *Polym. Degrad. Stab.* **2002**, *75*, 543–547. [CrossRef]
4. Chen, J.; Zhang, J.; Zhu, T.; Hua, Z.; Chen, Q.; Yu, X. Blends of thermoplastic polyurethane and polyether–polyimide: Preparation and properties. *Polymer* **2001**, *42*, 1493–1500. [CrossRef]
5. Chen, X.; Ma, C.; Jiao, C. Enhancement of flame-retardant performance of thermoplastic polyurethane with the incorporation of aluminum hypophosphite and iron-graphene. *Polym. Degrad. Stab.* **2016**, *129*, 275–285. [CrossRef]
6. Lorenzetti, A.; Modesti, M.; Besco, S.; Hrelja, D.; Donadi, S. Influence of phosphorus valency on thermal behaviour of flame retarded polyurethane foams. *Polym. Degrad. Stab.* **2011**, *96*, 1455–1461. [CrossRef]
7. Zhao, Z.; Jin, Q.; Zhang, N.; Guo, X.; Yan, H. Preparation of a novel polysiloxane and its synergistic effect with ammonium polyphosphate on the flame retardancy of polypropylene. *Polym. Degrad. Stab.* **2018**, *150*, 73–85. [CrossRef]
8. Fan, Y.; Wang, G.; Huang, X.; Bu, J.; Sun, X.; Jiang, P. Molecular structures of (3-aminopropyl)trialkoxysilane on hydroxylated barium titanate nanoparticle surfaces induced by different solvents and their effect on electrical properties of barium titanate based polymer nanocomposites. *Appl. Surf. Sci.* **2016**, *364*, 798–807. [CrossRef]
9. Luo, H.; Rao, W.; Zhao, P.; Wang, L.; Liu, Y.; Yu, C. An efficient organic/inorganic phosphorus-nitrogen-silicon flame retardant towards low-flammability epoxy resin. *Polym. Degrad. Stab.* **2020**, *178*, 109195. [CrossRef]
10. Hu, X.; Yang, H.; Jiang, Y.; He, H.; Liu, H.; Huang, H.; Wan, C. Facile synthesis of a novel transparent hyperbranched phosphorous/ nitrogen-containing flame retardant and its application in reducing the fire hazard of epoxy resin. *J. Hazard. Mater.* **2019**, *379*, 120793. [CrossRef]
11. Duan, L.; Yang, H.; Song, L.; Hou, Y.; Wang, W.; Gui, Z.; Hu, Y. Hyperbranched phosphorus/nitrogen-containing polymer in combination with ammonium polyphosphate as a novel flame retardant system for polypropylene. *Polym. Degrad. Stab.* **2016**, *134*, 179–185. [CrossRef]
12. Jin, Q.F.; Liao, G.X.; Jian, X.G. Synthesis and characterization of trimethoxysilyl-functionalized poly(phthalazinone ether ketone). *Chin. Chem. Lett.* **2007**, *8*, 1137–1140. [CrossRef]
13. Paquet, O.; Salon, M.C.B.; Zeno, E.; Belgacem, M.N. Hydrolysis-condensation kinetics of 3-(2-amino-ethylamino)propyl-trimethoxysilane. *Mater. Sci. Eng. C* **2012**, *32*, 487–493. [CrossRef]
14. Velencoso, M.M.; Ramos, J.M.; Klein, R.; Lucas, A.D.; Rodriguez, F.J. Thermal degradation and fire behaviour of novel polyurethanes based on phosphate polyols. *Polym. Degrad. Stab.* **2014**, *101*, 40–51. [CrossRef]
15. Shi, Y.; Wang, G. The novel silicon-containing epoxy/PEPA phosphate flame retardantfor transparent intumescent fire resistant coating. *Appl. Surf. Sci.* **2016**, *385*, 453–463. [CrossRef]
16. Wu, C.S.; Liu, Y.L.; Chiu, Y.S. Epoxy resins possessing flame retardant elements from silicon incorporated epoxy compounds cured with phosphorus or nitrogen containing curing agents. *Polymer* **2002**, *43*, 4277–4284. [CrossRef]
17. Liu, W.; Chen, D.Q.; Wang, Y.Z.; Wang, D.Y.; Qu, M.H. Char-forming mechanism of a novel polymeric flame retardant with char agent. *Polym. Degrad. Stab.* **2007**, *92*, 1046–1052. [CrossRef]
18. Chen, X.; Hu, Y.; Jiao, C.; Song, L. Preparation and thermal properties of a novel flame-retardant coating. *Polym. Degrad. Stab.* **2007**, *92*, 1141–1150. [CrossRef]
19. Wang, Z.; Wei, P.; Qian, Y.; Liu, J. The synthesis of a novel graphene-based inorganic–organic hybrid flame retardant and its application in epoxy resin. *Composites B* **2014**, *60*, 341–349. [CrossRef]
20. Teng, C.C.; Ma, C.C.M.; Lu, C.H.; Yang, S.Y.; Lee, S.H.; Hsiao, M.C.; Yen, M.Y.; Chiou, K.C.; Lee, T.M. Thermal conductivity and structure of non-covalent functionalized graphene/epoxy composites. *Carbon* **2011**, *49*, 5107–5116. [CrossRef]

21. Zhou, S.; Song, L.; Wang, Z.; Hu, Y.; Xing, W. Flame retardation and char formation mechanism of intumescent flame retarded polypropylene composites containing melamine phosphate and pentaerythritol phosphate. *Polym. Degrad. Stab.* **2008**, *93*, 1799–1806. [CrossRef]
22. Dutkiewicz, M.; Przybylak, M.; Januszewski, R.; Maciejewski, H. Synthesis and flame retardant efficacy of hexakis(3-(triethoxysilyl)propyloxy)cyclotriphosphazene/silica coatings for cotton fabrics. *Polym. Degrad. Stab.* **2018**, *148*, 10–18. [CrossRef]
23. Kim, H.S.; Kim, H.J. Enhanced hydrolysis resistance of biodegradable polymers and bio-composites. *Polym. Degrad. Stab.* **2008**, *93*, 1544–1553. [CrossRef]
24. Kannan, G.A.; Choudhury, N.R.; Dutta, K.N. Synthesis and characterization of methacrylate phospho-silicate hybrid for thin film applications. *Polymer* **2007**, *48*, 7078–7086. [CrossRef]
25. Chana, M.L.; Lau, K.T.; Wong, T.T.; Cardona, F. Interfacial bonding characteristic of nanoclay/polymer composites. *Appl. Surf. Sci.* **2011**, *258*, 860–864. [CrossRef]
26. Chai, B.; Yan, J.; Wang, C.; Ren, Z.; Zhu, Y. Enhanced visible light photocatalytic degradation of Rhodamine B over phosphorus doped graphitic carbon nitride. *Appl. Surf. Sci.* **2017**, *391*, 376–383. [CrossRef]
27. Zhang, W.; Li, X.; Fan, H.; Yang, R. Study on mechanism of phosphorus–silicon synergistic flame retardancy on epoxy resins. *Polym. Degrad. Stab.* **2012**, *97*, 2241–2248. [CrossRef]
28. Zhang, Y.; Ni, Y.P.; He, M.X.; Wang, X.L.; Chen, L.; Wang, Y.Z. Phosphorus-containing copolyesters: The effect of ionic group and its analogous phosphorus heterocycles on their flame-retardant and anti-dripping performances. *Polymer* **2015**, *60*, 50–61. [CrossRef]
29. Huang, C.; Puziy, M.A.; Sun, T.; Poddubnaya, I.O.; Fabián, S.G.; Juan, M.D.T.; Denisa, H.J. Capacitive behaviours of phosphorus-rich carbons derived from lignocelluloses. *Electrochim. Acta* **2014**, *137*, 219–227. [CrossRef]
30. Brookes, N.P.; Fraser, S.; Short, D.R.; Hanley, L.; Fuoco, E.; Roberts, A.; Hutton, S. The effect of ion energy on the chemistry of air-aged polymer films grown from the hyperthermal polyatomic ion Si2OMe5+. *J. Electron. Spectrosc. Relat. Phenom.* **2001**, *121*, 281–297. [CrossRef]
31. Wang, X.; Wu, L.; Li, J. Synergistic flame retarded poly(methyl methacrylate) by nano-ZrO2 and triphenylphosphate. *J. Therm. Anal. Calorim.* **2011**, *103*, 741–746. [CrossRef]

sustainability

MDPI

Article

Utilization of Silicon Carbide Sludge as Metakaolin-Based Geopolymer Materials

Kae-Long Lin [1], Kang-Wei Lo [2,3,*], Ta-Wui Cheng [3], Wei-Ting Lin [4] and Ya-Wen Lin [3]

1. Department of Environmental Engineering, National Ilan University, Ilan City 260, Taiwan; kllin@niu.edu.tw
2. Graduate Institute of Engineering Technology, National Taipei University of Technology, Taipei City 106, Taiwan
3. Institute of Mineral Resources Engineering, National Taipei University of Technology, Taipei City 106, Taiwan; twcheng@ntut.edu.tw (T.-W.C.); urine1001@gmail.com (Y.-W.L.)
4. Department of Civil Engineering, National Ilan University, Ilan 260, Taiwan; wtlin@niu.edu.tw

* Correspondence: dark83054689@gmail.com; Tel.: +886-3935-7579

Received: 10 August 2020; Accepted: 3 September 2020; Published: 7 September 2020

Abstract: The recycling of SiC sludge material is crucial for resource reutilization and environmental protection. In the current study, the effect of the mass ratio between the Na_2SiO_3 and sodium hydroxide (NaOH) solutions (NS/SS ratio) and the effect of SiC sludge on metakaolin geopolymers was comprehensively investigated to determine the underlying performance of the geopolymerization system. During thermal evolution, the second exothermic peak of 1.6NS10SCS (NS/SS ratio: 1.6, 10% SiC sludge) showed a heat evolution value of 990.6 W/g, which was the highest among other geopolymers. Additionally, the 1.6NS10SCS sample after 28 days of curing showed the highest flexural strength (6.42 MPa), compared to that of the others, and the DTA/TG (differential thermal analysis/thermogravimetry) results showed that the weight loss percentage increased to 14.62% from 400 to 750 °C. For the ^{29}Si nuclear magnetic resonance deconvolution, 1.6NS10SCS exhibited high fractions of Q^4(3Al) (33.63%), Q^4(2Al) (23.92%), and Q^4(1Al) (9.70%). Thus, the geopolymer with the optimal SiC-sludge replacement level and NS/SS ratio contained more macropores and geopolymer gels, which benefit structural development. The experimental results indicated that SiC-sludge can potentially serve as a partial replacement for metakaolin and exhibited favorable mechanic characteristics.

Keywords: utilization; SiC sludge; alkaline activator solutions; synergistic effect; geopolymer reaction

1. Introduction

The solar cells generated electricity will become the primary global energy source, because the solar power was safe, efficient, nonpolluting and reliable [1]. Recycling silicon is of great importance, because the demand for silicon for solar cells is increasing globally, while concern about the limited silicon supply is emerging [2]. Generally, solar cells use multicrystalline wafers made of silicon [1]. Crystalline silicon cells have laboratory energy conversion efficiencies over 25% for single-crystal cells and over 20% for multicrystalline cells. In 2008, more than 100 GW silicon solar panels were produced based on an energy-intensive Siemens process [3], which cost approximately 400,000 tons of silicon. There were various types of solar cells, whereby the C-Si solar cell dominates 80% of the market globally [1]. Industry mainly uses silicon carbide (SiC) for wafer slicing [4], and a substantial amount of SiC sludge is formed when a silicon ingot is cut into wafers, with a kerf loss of over 50% [5]. The kerf-loss silicon waste is greater than 200,000 tons per year, and this sludge contributes to environmental pollution. Therefore, the recycling of SiC sludge material is crucial for resource reutilization and environmental protection. In addition, the primary composition of SiC sludge is SiO_2 and Al_2O_3. Therefore, SiC sludge can be used as a raw material in producing silicon carbide sludge metakaolin-based (SCSMB) geopolymers.

Inorganic polymer (geopolymer) is a cementation material that can be used as an alternative to ordinary Portland cement. Geopolymers are mostly formed by mixing silicate aluminum sources (industrial waste such as slag and fly ash) with alkaline silicate solutions or an alkaline metal [6]. The polymerization process produced a three-dimensional (3D) compound comprising of [AlO_4] and [SiO_4] tetrahedrons [7,8]. Geopolymerization begins with the generation of -Si-O-Al-O-Si- or -Si-O-Al- monomers (oligomers) in the solution. These monomers join to form a geopolymer. As a structural material, a geopolymer is a potential replacement material for ordinary Portland cement (OPC). The most common alkaline activator solution (AAS) is a combination of Na_2SiO_3 and NaOH. AAS plays a crucial role in the polymerization process, which in turn has a strong dependence on the Si/Al ratio, M_2O-H_2O ratio (where M = Na and/or K), hydroxide concentration, and alkali cation used [9–12].

Recently, there have been many studies on SiC sludge applied in the field of geopolymer materials. For example, Medri et al. [13] reported that silicon carbide (SiC) foams were developed using a low-temperature process, such as chemical consolidation, that is suitable for replacing the sintering step. The amorphous silica layer covering the surface of SiC particles participates in geopolymerization so that nanoprecipitates easily form on the SiC surface [13]. Rahman et al. [14] reported that silicon carbide whiskers were introduced into sodium-based geopolymers in order to evaluate the response of silicon carbide whiskers to the interfacial bonding and strength of sodium-based geopolymers along with rice husk ashes. Results show that the simultaneous additions of rice husk ashes and silicon carbide whiskers resulted in flexural strength improvement by ~27% and ~97%, respectively. The increase in flexural strength due to simultaneous inclusion of silica and silicon carbide whiskers suggests that silica particles are compatible with the metakaolin-based geopolymers, which is effective in consolidation [14]. Medri et al. [14] reported that refractory paints based on silicon carbide (SiC) were developed using inorganic alkali-aluminosilicate binders. R-SiC coatings applied by brushing on a Si_3N_4-TiN substrate cannot act as barriers against oxygen penetration due to the presence of large pores [15]. Bai et al. [16] reported that high-strength open-celled SiC/geopolymer foam composites (SFCs) were fabricated by a combined saponification/peroxide route. The optimal sample possessed a total porosity of up to ~84 vol%, an open porosity of ~83 vol%, a thermal conductivity of 0.15 W/mK, and a compression strength of ~1.1 MPa [16]. Jia et al. [17] reported that bulk and porous AlN/SiC-based ceramics were prepared in situ from graphite/geopolymer composites based on a carbon thermal reduction reaction under high temperatures. Moreover, both h-AlN and SiC ceramics derived by this technique show porous features, with porosities of 61% and 68% and considerable compressive strength values of 0.76 MPa and 0.63 MPa, respectively [17]. Although, previous investigations have studied SiC sludge in the field of geopolymer materials, such as silicon carbide (SiC) foams [13], SiCf-reinforced geopolymers [14], refractory paints [15], high-strength open-celled SiC/geopolymer foam composites (SFCs) [16], and porous AlN/SiC-based ceramics [17].

Most of the previous investigations were based on studies on SiC sludge applied in the field of geopolymer materials, which led to knowledge; few reports are available regarding the effect of alkaline-activator solution on the geopolymerization of SiC sludge-based geopolymers. In this study, a new way to the effect of the mass ratio between the Na_2SiO_3 and sodium hydroxide (NaOH) solutions (NS/SS ratio) and the effect of SiC sludge on metakaolin geopolymers was comprehensively investigated to determine the underlying performance of the geopolymerization system. The flexural strengths of the geopolymers were examined in a laboratory, and their microstructures were characterized using an isothermal conduction calorimeter, differential thermal analysis (DTA/TG), and nuclear magnetic resonance (NMR).

2. Materials and Methods

The silicate-aluminum sources were metakaolin (MK) and SiC sludge. The SiC sludge was gained from a plant that makes LED substrates (Taiwan). The major chemical composition of the SiC sludge was 75.4 wt % SiO_2, 23 wt % SiC, and 0.8 wt % Al_2O_3; the major chemical composition of the MK was 51.8 wt % SiO_2 and 43 wt % Al_2O_3, as shown in Table 1. Sodium hydroxide (NaOH, 10 M) and distilled water were mixed with a commercial sodium silicate solution (Na_2SiO_3), Ms (SiO_2 mole/Na_2O mole) = 3.1 (SiO_2 = 28.1%, Na_2O = 9.09%, H_2O = 62.8%). The mass ratio between Na_2SiO_3 and sodium hydroxide (NaOH) solutions (NS/SS ratio) was in the range of 0.8 to 2.0 (0.8, 1.2, 1.6, and 2.0) and analysis of the effect of an alkaline-activator solution on a geopolymer reaction. Calculate the amount of NS/SS ratio as follows.

NS/SS ratio = SiO_2 mole (Sodium silicate solution)/[Na_2O mole (Sodium silicate solution) + Na_2O mole (10 M NaOH)]

Table 1. The major chemical composition of materials.

Composition	SiC Sludge	MK
SiO_2 (wt %)	75.40	51.80
Al_2O_3 (wt %)	0.80	43.00
Fe_2O_3 (wt %)	0.58	1.30
CaO (wt %)	0.09	0.25
MgO (wt %)	N.D. [1]	N.D. [1]
SO_3 (wt %)	0.06	N.D. [1]
Na_2O (wt %)	N.D. [1]	0.04
K_2O (wt %)	0.01	0.32
SiC (wt %)	23.00	N.D. [1]

[1] N.D.: not detected.

Then, MK and 0–40 wt % SiC sludge were added into an alkali activator, and analysis of the effect of a Si/Al mole ratios on a geopolymer reaction. The ratios of material in the mixture of geopolymers are listed in Table 2. Calculate the amount of Si/Al mole ratios as follows. First, the MK, SiC sludge, and alkaline solution were stirred for 60 min, and deionized water (3.8 L) was subsequently added to the blend to generate a suspension by stirring for 30 s. After filtration, the dissolved Si and Al concentrations were measured by atomic absorption spectroscopy (AAS). SiC-sludge metakaolin-based geopolymer samples were synthesized by mixing powder and alkaline activator solution for 5 min, and the paste was poured into a plastic mold for the analysis of microstructure characteristics and mechanical properties. The flexural strength tests were performed after 1, 7, 14, 28, and 56 days using a Hung Ta HT-2402 testing machine (Hung Ta Instruments, Taichung City, Taiwan) with a three-point bending test method, according to ASTM C348. The average strength value of the three specimens is presented, and the coefficient of variation of these results is less than 10%. The mechanical properties of the three specimens were evaluated and expressed. Finally, the samples were analyzed at appropriate curing times by DTA/TG and ^{29}Si MAS NMR analyses. The heat evolution in the geopolymerization reactions, such as dissolution, polymerization, and condensation, were characterized by an isothermal conduction calorimeter (Calmetrix I-CAL 2000 HPC, Boston, MA, USA) conducted at a constant temperature of 30 ± 0.02 °C for 168 h, according to ASTM C1702. The DTA/TG studies were conducted, in which the heating range was from 50 °C to 1000 °C by an STA6000 thermal analyzer. High-resolution 29Si MAS/NMR spectra were recorded at 39.72 MHz on an MSL Bruker MAS/NMR-200 solid-state high-resolution spectrometer using rapid (approximately 3 kHz) sample spinning at the magic angle to the external magnetic field. Analysis method based on reference [18], the coordination of Q^4(mAl) species (4Al, 3Al, 2Al, 1Al, 0Al) in the SCSMB geopolymer was obtained by applying Seasolve PeakFit™ software, version 4 (Seasolve Software Inc., Framingham, MA, USA).

Table 2. The ratios of material in mixture.

Mix Designation	NS/SS	MK (g)	SiC Sludge (g)	Sodium Silicate Solution (g)	10 M NaOH (g)	Si/Al Mole Ratios
0.8NS0SCS		500	0	240	260	1.01
0.8NS10SCS		450	50	240	260	1.36
0.8NS20SCS	0.8	400	100	240	260	1.24
0.8NS30SCS		350	150	240	260	1.74
0.8NS40SCS		300	200	240	260	2.60
1.2NS0SCS		500	0	315	185	0.98
1.2NS10SCS		450	50	315	185	1.34
1.2NS20SCS	1.2	400	100	315	185	1.22
1.2NS30SCS		350	150	315	185	1.87
1.2NS40SCS		300	200	315	185	2.72
1.6NS0SCS		500	0	374	126	1.16
1.6NS10SCS		450	50	374	126	1.51
1.6NS20SCS	1.6	400	100	374	126	1.43
1.6NS30SCS		350	150	374	126	1.92
1.6NS40SCS		300	200	374	126	2.89
2.0NS0SCS		500	0	421	79	1.07
2.0NS10SCS		450	50	421	79	1.48
2.0NS20SCS	2.0	400	100	421	79	1.23
2.0NS30SCS		350	150	421	79	1.78
2.0NS40SCS		300	200	421	79	2.67

3. Results

3.1. Heat Evolution of Geopolymers with SiC Sludge

Geopolymers comprise a framework containing inorganic T-O (T: Si or Al). Geopolymerization is analogous to the curing process of polymers, which includes gelation, polymerization, and solidification [9]. The heat evolution can be measured by an isothermal conduction calorimeter, which indicates the progress of the geopolymerization reaction. In all cases, the first exothermic peak, which is related to the dissolution of solid materials in the alkaline activator, is quite sharp at the beginning of the preinduction period [19]. The samples were mixed outside the device, and the heat evolution was so fast that there was not sufficient time to achieve thermal equilibrium; thus, the first peak was only partially registered. The second exothermic peak indicated continuous dissolution and polymerization, during which oligomers combined to form a geopolymer network.

Figure 1a–d illustrates the effects of the NS/SS ratio on the heat evolution during geopolymerization reactions. When the NS/SS ratio was 0.8, the heat evolutions of 0.8NS0SCS, 0.8NS10SCS, 0.8NS20SCS, and 0.8NS30SCS were 806.5, 720.3, 415.3, and 445.8 W/g heat, respectively (Figure 1a). The high NaOH concentration prompted the dissolution of the initial solid in the alkaline activator, which caused the first exothermic peaks of 0.8NS0SCS and 0.8NS10SCS to be more intense than those of the others. In addition, the second exothermic peaks of 0.8NS0SCS and 0.8NS10SCS appeared at approximately 6.65 h and 9.98 h, respectively. This indicated that a higher NaOH content promoted the reactivity of the geopolymerization reaction. However, the second exothermic peak of 0.8NS40SCS emerged at 14.82 h. According to Hu et al. [20], the geopolymer gel formed by the rapid reaction of active aluminosilicate under a high concentration of alkali hindered the subsequent dissolution of the FA (Fly ash) [20]. The results indicate that the amounts of dissolved silica and alumina from MK are limited due to precipitations of geopolymer gels around the surface of SCS particles. Therefore, the heat evolution duration of the second exothermal peak clearly increased from 6.65 h to 14.82 h as the SiC sludge-replacement levels increased from 0% to 40%, as shown in Figure 1a. An increase in the NS/SS ratio (1.2) was expected to improve the geopolymerization reaction, and the results of 1.2NS10SCS showed that the second exothermic peak reached 958.3 W/g. A synergistic effect between the SCS and the MK was suggested to contribute to the promotion of the geopolymerization reaction, which caused the heat evolution of 1.2NS10SCS to be higher than that of 1.2NS0SCS (945.2 W/g). Additionally, the heat evolution duration of the second exothermal peak increased from 14.82 to 17.82 h as the NS/SS ratio increased from 0.8 to 1.2 (0.8NS40SCS and 1.2NS40SCS). According to Figure 1b, it can be inferred that, although the NS/SS ratio increased, the amounts of dissolved silica and alumina from MK were limited due to the precipitations of geopolymer gels around the surface of SCS (SiC sludge) particles.

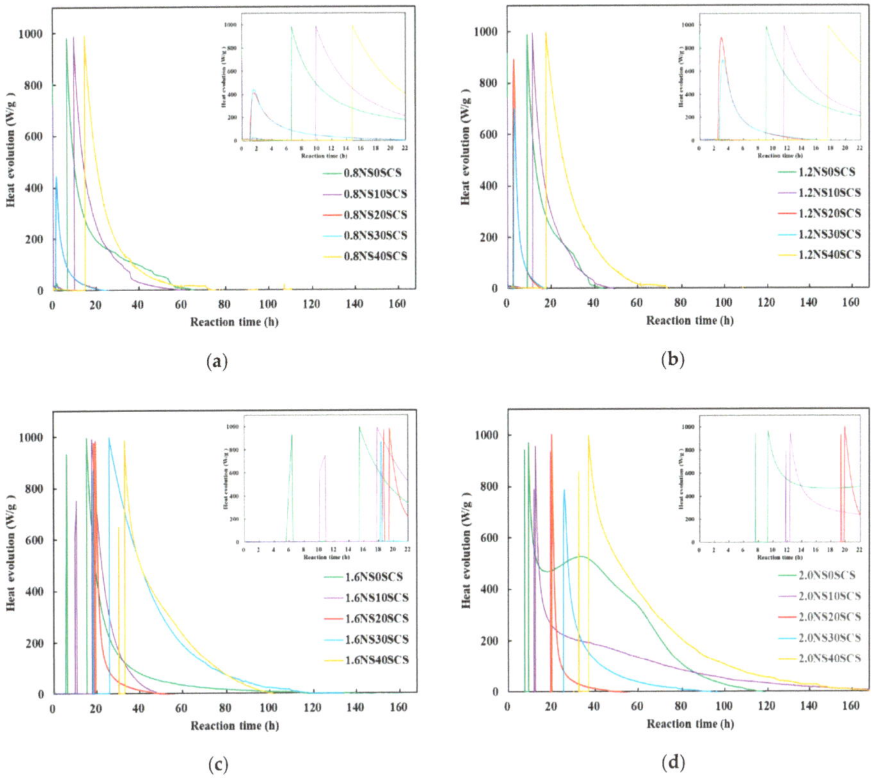

Figure 1. Effects of NS/SS (Na_2SiO_3/NaOH solutions) ratios on heat evolution rate for geopolymer with SiC sludge: (**a**) NS/SS ratio = 0.8; (**b**) NS/SS ratio = 1.2; (**c**) NS/SS ratio = 1.6; (**d**) NS/SS ratio = 2.0.

Figure 1c–d shows the heat evolution curves of the SCSMB geopolymer sample with NS/SS values of 1.6 and 2.0, in which two exothermic peaks are classified. The heat evolution (990.6 W/g) of the second exothermic peak of 1.6NS10SCS was higher than those of all geopolymers. This fact implies that the synergistic effect promoted the oligomers to combine and form integral geopolymer gels [20]. However, for 2.0NS10SCS, the heat evolution of the second exothermic peak decreased from 990.6 W/g to 951.5 W/g with an increase in the NS/SS ratio. According to Lo et al. [21], decreasing the concentration of Na_2O caused the decreased of the amount of dissolved Si and Al and restrained the geopolymer synthesis reaction.

3.2. Mechanical Properties

The results of the flexural strength of the geopolymer with SiC sludge cured for 1–56 days are shown in Figure 2. The flexural strength of geopolymers varied at different replacement levels of SiC sludge and was positively correlated with the replacement levels. On the first day of the curing process, the flexural strengths of 0.8NS0SCS, 0.8NS10SCS, 0.8NS20SCS, 0.8NS30SCS, and 0.8NS40SCS were 4.12, 3.97, 3.89, 3.56, and 3.57 MPa, respectively (Figure 2a). The amounts of dissolved silica and alumina from MK are limited due to geopolymer gel precipitations around the surface of SCS particles. This reduction generated an inadequate amount of leading precursors, thus reducing the strength. Hu et al. [20] discussed that the decrease in strength at a longer curing age might be due to depolymerization or dissolution of aluminosilicates resulting from the low SiO_2/Al_2O_3 molar ratio in formulation, which was 2.15 for the G-RM-5N-0FA. In particular, the strength of the G-RM-5N-0FA cured for 7 days was below 5 MPa [20]. The flexural strength for the 28 day SiC sludge geopolymers is 4.05 MPa for 0.8NS30SCS (SiO_2/Al_2O_3 molar ratio of 1.74), as shown in Figure 2a. A low SiO_2/Al_2O_3 molar ratio significantly reduced the strength for 0.8NS30SCS, which is in line with the study from Hu et al. for geopolymer materials [20].

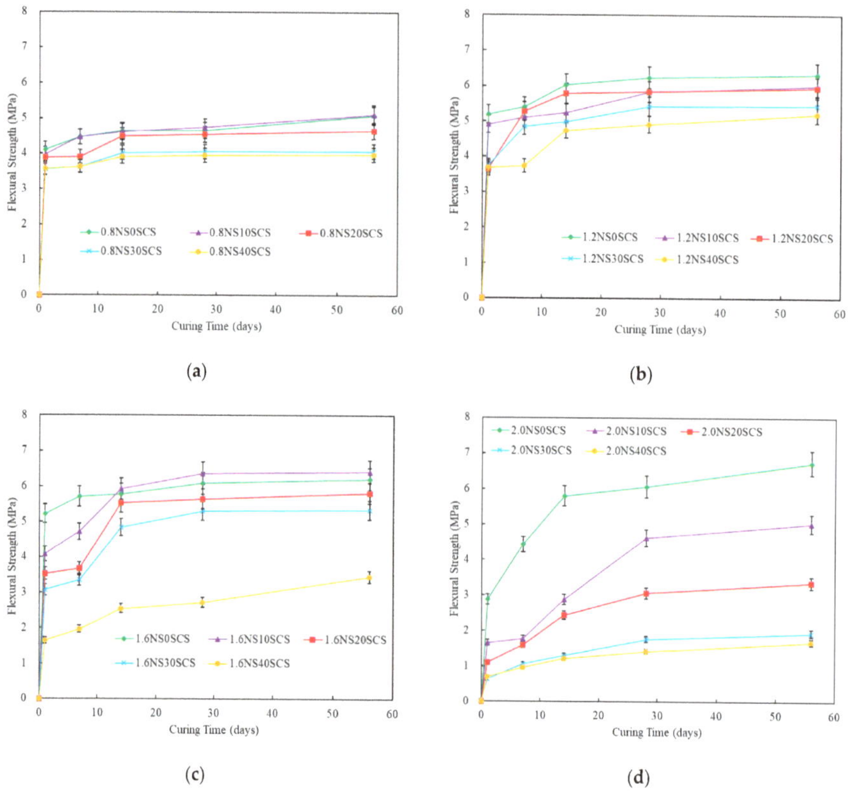

Figure 2. Flexural strength of geopolymer with SiC sludge cured for 1–56 days: (**a**) NS/SS ratio = 0.8; (**b**) NS/SS ratio = 1.2; (**c**) NS/SS ratio = 1.6; (**d**) NS/SS ratio = 2.0.

For the NS/SS ratio of 1.2, the flexural strengths of 1.2NS20SCS, 1.2NS30SCS, and 1.2NS40SCS steadily decreased to 3.64, 3.74, and 3.69 MPa on the first day of curing, respectively (Figure 2b). It is possibly the increase in the NaOH content, which generated a sufficient amount of "OH^-" in the system and thus led to the increased solubility. This behavior caused unsatisfactory polycondensation and reduced the formation of amorphously structured geopolymer gels. The high Na_2O concentration may block the acceleration effect of geopolymer synthesis, leading to slow strength development, which is in line with previous studies for geopolymer materials [21].

In contrast, the increase in the curing time from 1 to 56 days increased the strengths of 1.6NS10SCS and 2.0NS10SCS from 4.08 to 6.42 MPa and from 1.66 to 5.03 MPa, respectively (Figure 2c). Increasing the NS/SS ratio promoted the dissolution of the initial solid, which accelerated the geopolymerization reaction and led to the rapid formation of geopolymer network structures. Therefore, the strength increased when the amount of amorphously structured geopolymer gel formed in the geopolymerization system increased. Lo et al. [21] reported that MKGP (MK-based geopolymer) with a 10% TLWD (TFT-LCD waste glass) replacement level showed the strength of 6.5 MPa with a S/N ratio of 1.6.

The results indicated that strength increased with the NS/SS ratio. Yaghoubi et al. [22] noted that the geopolymer strength was related to the degree of polycondensation, which was strongly influenced by the soluble SiO_4 and AlO_4 content of the polymer system. A high degree of polycondensation in the Si-O-Al framework structure corresponds to high strength [23]. In addition, the flexural strength of 1.6NS10SCS (SiO_2/Al_2O_3 molar ratio of 1.51) is also very high, as shown in Figure 2c. This increase in strength could be from the increased SiO_2/Al_2O_3 molar ratio. The SiO_2/Al_2O_3 molar ratio has a definite effect on strength, increasing the strength to 6.37 MPa. In addition, Rahman et al. [14] also noted that the increase in flexural strength due to simultaneous inclusion of silica and silicon carbide whiskers indicates that it is possible that silica particles are compatible with the metakaolin-based geopolymers, which is effective in consolidation [14]. However, the flexural strength of 1.6NS40SCS with a curing age of 28 days decreased to 2.73 MPa (SiO_2/Al_2O_3 molar ratio of 2.89), as shown in Figure 2c. With increasing SiC sludge content in the geopolymer, the SiO_2/Al_2O_3 molar ratio increases, but the amounts of dissolved silica and alumina from MK are limited due to precipitation of geopolymer gels around the surface of SCS particles. A greater amount of SiC sludge (over 20%) may block the synergistic effect of SCS and MK. Hu et al. [20] reported about when the FA (fly ash) content was 100 wt %, and the strength of the FA-based geopolymer with a curing age of 28 days was only 5 MPa (SiO_2/Al_2O_3 molar ratio of 4.38).

3.3. DTA/TG of Geopolymers with SiC Sludge

Figure 3 displays the effects of the geopolymers with SiC sludge on the DTA/TG results. According to the weight loss of the samples during heating, all the processes of DTA/TG results are categorized into Stages I, II, and III. The first peak at approximately 230 °C was associated with the loss of carbonate salt. Almost all the weight loss at the second peak, which occurred from 230 °C to 400 °C, was likely due to the dehydration of the macropore structure of the material [21]. The third peak, which occurred from 400 °C to 750 °C, was mainly related to the loss of the amorphous composition and dehydration of the geopolymer gel [21,24]. On the first day of curing, the weight loss percentages of 0.8NS0SCS, 0.8NS10SCS, 0.8NS20SCS, and 0.8NS30SCS were 10.21, 10.20, 10.09, and 10.08 wt %, respectively (Table 3). The amounts of dissolved silica and alumina from MK are limited due to precipitation of geopolymer gels around the surface of SCS particles, thus reducing the macropore structure. Therefore, the weight loss of 0.8NS40SCS was 10.06 wt %. The geopolymers with high replacement levels of SiC sludge generated an inadequate amount of leading precursors, thus limiting the polycondensation process. Therefore, the weight loss percentage of 0.8NS40SCS from 400 to 750 °C decreased to 1.95%, as shown in Table 3.

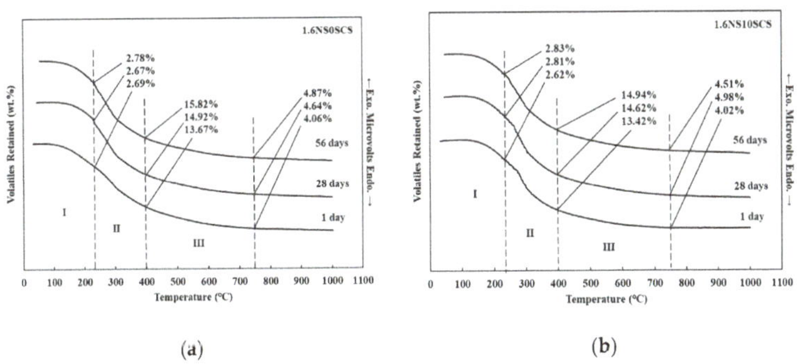

Figure 3. *Cont.*

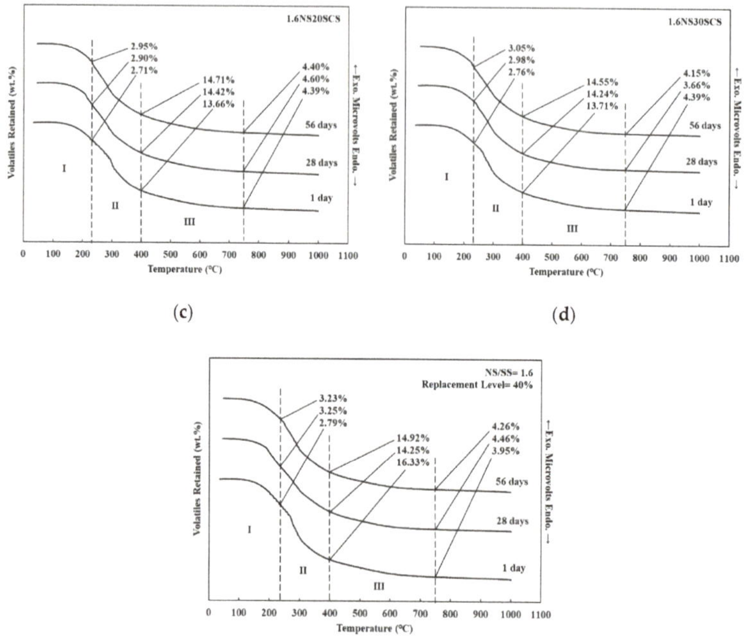

(c) (d)

(e)

Figure 3. The DTA/TG results of the geopolymers with SiC sludge and NS/SS ratio of 1.6, during heating: (**a**) replacement level = 0%; (**b**) replacement level = 10%; (**c**) replacement level = 20%; (**d**) replacement level = 30%; (**e**) replacement level = 40%.

When the NS/SS ratio increased from 0.8 to 1.6, the weight loss percentage of 1.6NS10SCS was 13.42% in a temperature range of 230–400 °C, as depicted in Figure 3b. The high NS/SS ratio prompted the dissolution of the initial solid and consequently promoted the geopolymerization reaction, which caused the rapid formation of 3D silico–aluminate structures. Furthermore, the synergistic effect between the SCS and the MK was assigned to the promotion of the geopolymerization reaction, which caused the oligomers to form integral geopolymer gels. Therefore, the structural strength increased with an increase in the amount of amorphously structured geopolymeric gel in the geopolymerization system. According to the results, the weight loss percentage of 1.6NS10SCS increased to 14.94% after 56 days of curing. Furthermore, low weight loss occurred in the temperature range of 400–750 °C. The weight loss percentages of 1.6NS0SCS, 1.6NS10SCS, 1.6NS20SCS, 1.6NS30SCS, and 1.6NS40SCS were 4.87, 4.51, 4.40, 4.15, and 4.26 wt %, respectively (Figure 3a–e).

Table 3. Weight loss of the geopolymers with SiC sludge during heating from ambient temperature to 1000 °C (NS/SS ratio = 0.8 and 1.2).

Samples	NS/SS	Curing Time (Days)	Temperature (°C)		
			50–230	230–400	400–750
0.8NS0SCS	0.8	1	2.58	10.21	2.61
		28	3.70	11.81	4.24
		56	4.37	12.36	4.68
0.8NS10SCS		1	2.77	10.20	2.35
		28	3.83	11.69	4.24
		56	4.42	12.36	4.62
0.8NS20SCS		1	3.72	10.09	2.33
		28	3.83	11.66	4.03
		56	4.48	12.32	4.24
0.8NS30SCS		1	3.49	10.08	2.22
		28	3.94	11.64	3.68
		56	4.48	12.24	4.03
0.8NS40SCS		1	3.49	10.06	1.95
		28	3.94	11.59	3.52
		56	4.50	12.21	3.68
1.2NS0SCS	1.2	1	3.49	10.97	2.73
		28	3.94	12.51	4.09
		56	4.51	12.96	4.77
1.2NS10SCS		1	3.49	10.96	2.72
		28	4.14	12.49	4.07
		56	4.55	12.93	4.34
1.2NS20SCS		1	3.52	10.93	2.64
		28	4.21	12.46	3.99
		56	4.55	12.86	4.33
1.2NS30SCS		1	3.52	10.92	2.64
		28	4.25	12.42	3.96
		56	4.61	12.75	4.21
1.2NS40SCS		1	3.58	10.88	2.57
		28	4.25	12.39	3.90
		56	4.61	12.65	4.14

3.4. ^{29}Si NMR Spectra and Deconvolutions of Geopolymers with SiC Sludge

To further validate the evolution of chemical reactions in geopolymerization with SiC sludge, ^{29}Si NMR analyses were conducted on the geopolymer samples. The ^{29}Si NMR spectra analysis enables the short-range ordering and molecular structure in the geopolymerization reaction to be determined [25]. In the initial stage, geopolymerization involved the release of the silicate derivatives (Q^1, Q^2, and Q^3) in the solution by alkaline activation. Here, the resonances of Q^1, Q^2, and Q^3 resonate at approximately −79, −85, and −95 ppm, respectively, and the resonance of Q^4 is the same as the Q^4(0Al) of the geopolymer gel [26]. The detailed reaction mechanism is discussed in the following sections. The resonance of a Q^4(mAl) ($0 \leq m \leq 4$) center with the replacement of each aluminum by silicon is approximately −5 ppm, with chemical shifts at −84, −89, −93, −99, and −107 ppm that correspond to Q^4(4Al), Q^4(3Al), Q^4(2Al), Q^4(1Al), and Q^4(0Al), respectively [25].

Figure 4 displays the ^{29}Si NMR spectra and deconvolutions of SiC sludge geopolymers with 28 days of curing. After 28 days of curing, the spectra of 0.8NS0SCS revealed a high percentage of Q^4(3Al), thus indicating a high initial level of aluminosilicate. For the ^{29}Si NMR deconvolution of 0.8NS0SCS, the percentages of Q^4(3Al), Q^4(2Al), and Q^4(1Al) were 34.32%, 16.06%, and 7.06%, respectively. However, geopolymers with high replacement levels of SiC sludge generated an inadequate amount of leading precursors, thus limiting the polycondensation process. For example, for 0.8NS40SCS, the percentages of Q^4(3Al), Q^4(2Al), and Q^4(1Al) were 29.59%, 24.91%, and 20.05%, respectively. On the 28th day of curing, the spectra of 0.8NS40SCS revealed a high percentage of Q^4(0Al), thus indicating the formation of sodium silicate glass from the unreacted Na_2SiO_3 activator [27]. This result indicates that SiC sludge inhibits geopolymerization reactions.

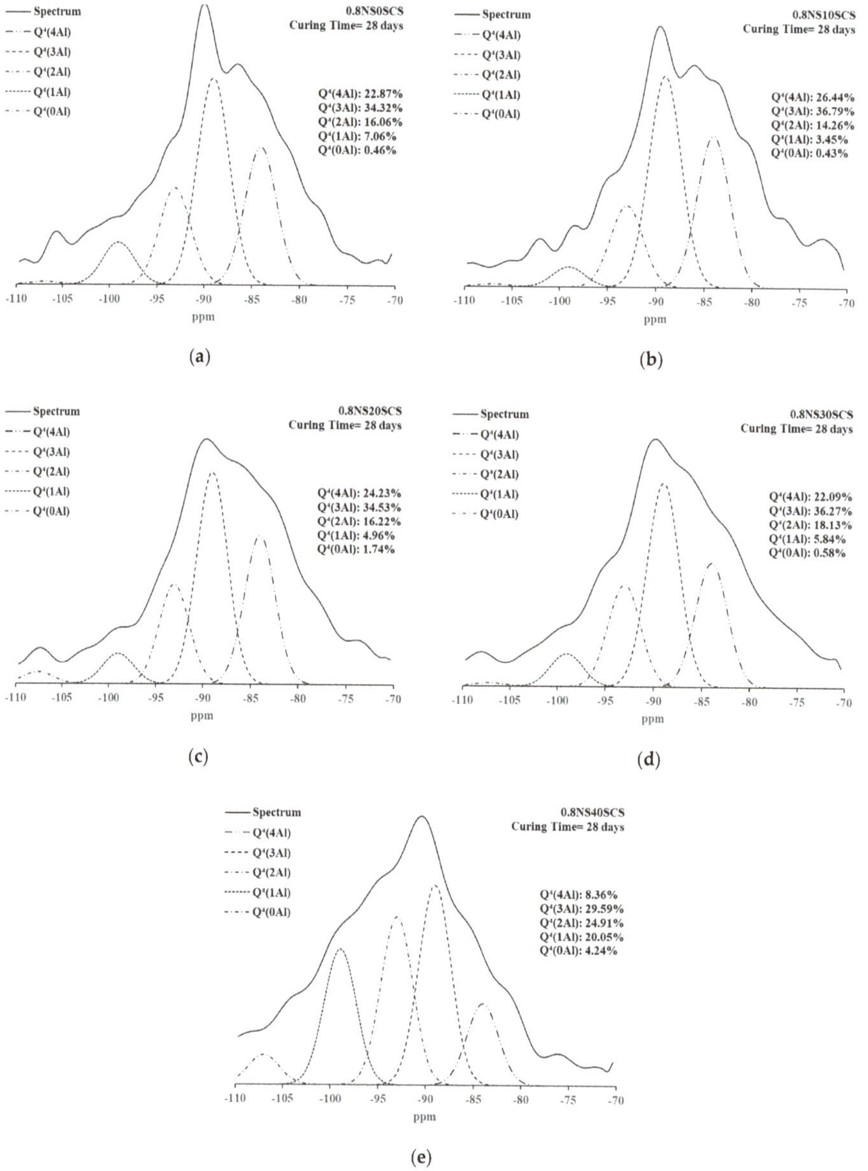

Figure 4. The ^{29}Si NMR spectra and deconvolutions of SiC sludge geopolymers at Curing Time = 28 day, and NS/SS ratio of 0.8: (**a**) replacement level = 0%; (**b**) replacement level = 10%; (**c**) replacement level = 20%; (**d**) replacement level = 30%; (**e**) replacement level = 40%.

Figure 5 depicts the ^{29}Si NMR spectra deconvolution obtained for the geopolymers with SiC sludge over 28 days of curing. After the 28th day of curing for 1.2NS10SCS, the percentages of Q^4(3Al), Q^4(2Al), and Q^4(1Al) were 33.71%, 22.47%, and 9.03%, respectively. The results revealed that when the NS/SS ratio was relatively low, an increase in the NaOH content caused the generation of a sufficient amount of OH^- in the system to increase the solubility, thus limiting the polycondensation process. This behavior caused a decrease in the formation of amorphously structured geopolymer gels. Therefore, the Q^4(3Al), Q^4(2Al), and Q^4(1Al) contents of 1.2NS10SCS slowly increased, as displayed in Figure 5b. When the NS/SS ratio increased to 1.6, 1.6NS10SCS exhibited high percentages of Q^4(3Al) (33.63%), Q^4(2Al) (23.92%), and Q^4(1Al) (9.70%), as illustrated in Figure 6b. Because of a synergistic effect between silicon carbide sludge and metakaolin, the silica and alumina of metakaolin were dominant in the geopolymerization reaction, and the addition of silicon carbide sludge, which had silica was provided more reaction paths. Wan et al. [28] indicated that geopolymers with high compressive strengths have high percentages of Q^4(3Al), Q^4(2Al), and Q^4(1Al), and were possibly N–A–S–H gels [28]. The percentages of Q^4(3Al, 2Al, and 1Al) silicon centers in the ^{29}Si NMR spectra increased, thus indicating the formation of an increased number of tetrahedral aluminum-linking sites with tetrahedral silicon (Figure 6).

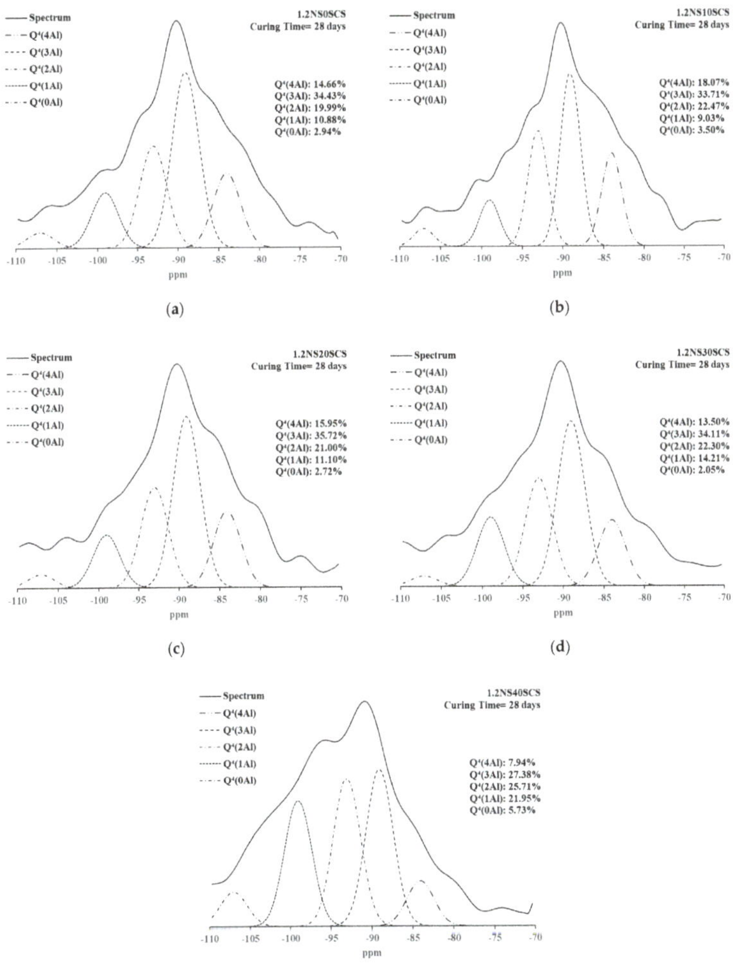

Figure 5. The ^{29}Si NMR spectra and deconvolutions of SiC sludge geopolymers at Curing Time = 28 day, and NS/SS ratio of 1.2: (**a**) replacement level = 0%; (**b**) replacement level = 10%; (**c**) replacement level = 20%; (**d**) replacement level = 30%; (**e**) replacement level = 40%.

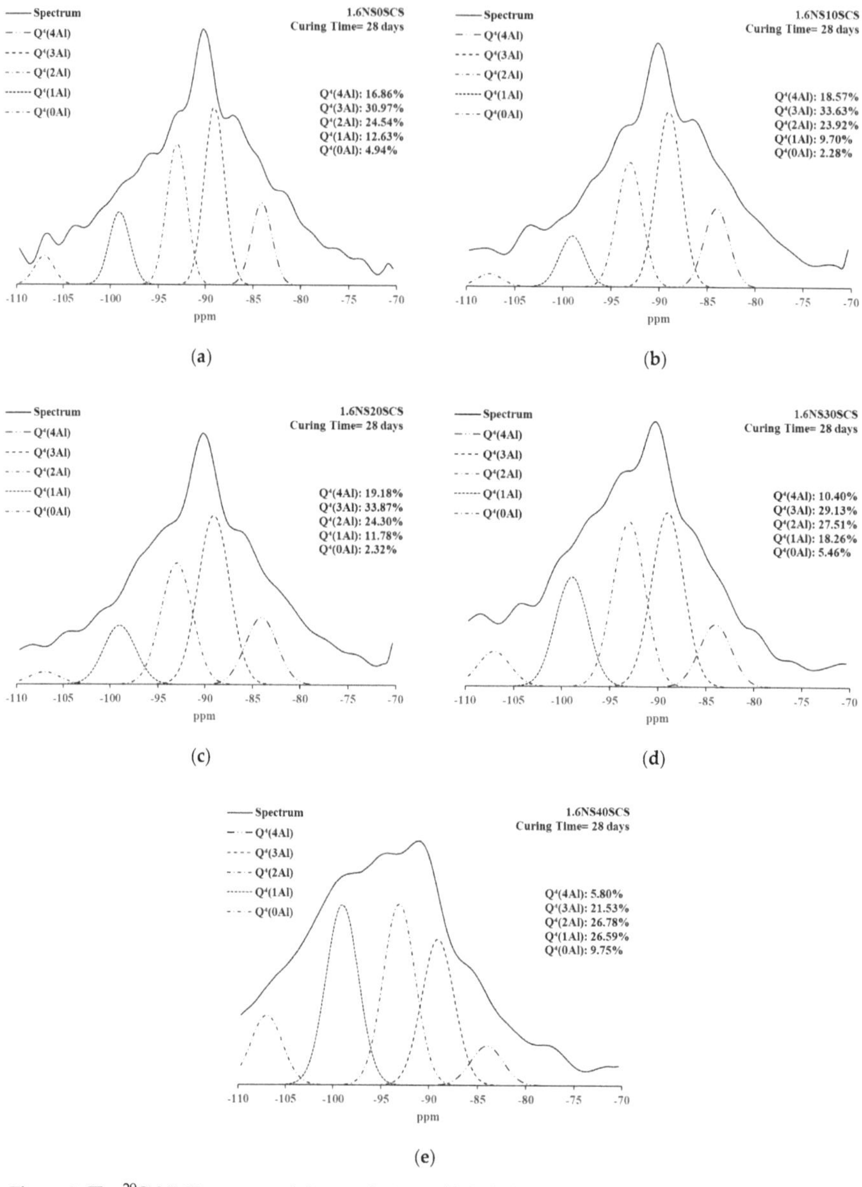

Figure 6. The ^{29}Si NMR spectra and deconvolutions of SiC sludge geopolymers at Curing Time = 28 day, and NS/SS ratio of 1.6: (**a**) replacement level = 0%; (**b**) replacement level = 10%; (**c**) replacement level = 20%; (**d**) replacement level = 30%; (**e**) replacement level = 40%.

4. Discussion

The geopolymerization process of geopolymers with SiC sludge is illustrated in Figure 7. In steps 1 and 2, the active components of metakaolin and SiC sludge particles were dissolved to form monosilicates (Q^0) that were linked to the end groups (Q^1), middle chain groups (Q^2), layers and branching sites (Q^3), and [SiO_4] and [AlO_4] tetrahedrons in the AASs. In step 3, primary particles of oligomeric gel were formed by the oligomers through the polycondensation reaction in the AASs. In step 4, the oligomers were polycondensed to form geopolymerization gels. In step 5, the polymers were polycondensed and accumulated around the unreacted SiC sludge surface to form 3D networks (Q^4).

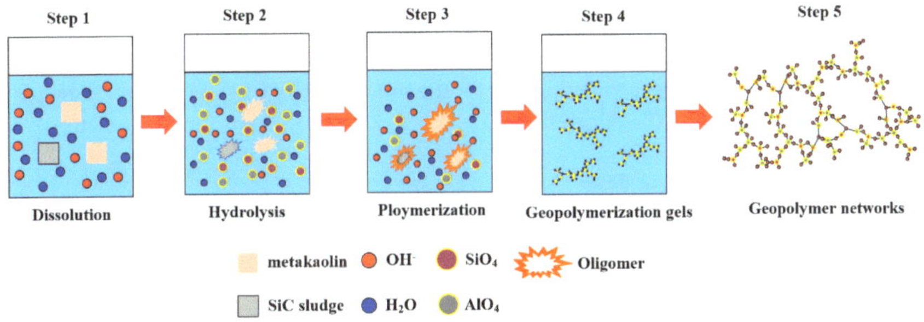

Figure 7. The geopolymerization process of geopolymers with SiC sludge.

At an early curing time, for 0.8NS0SCS, the sum of the silicate derivatives (Q^1, Q^2, and Q^3) was 13.59% (Table 4). After 28 days of curing, the percentage of silicate derivatives increased to 19.23%, as shown in Table 4. The results revealed that when the NS/SS ratio was relatively low, an increase in the NaOH content caused the generation of sufficient OH^- in the system to increase the solubility. This solubility increase caused hindrance in polycondensation, as indicated by the heat evolution rate and DTA/TG results (Figure 1 and Table 3). The inhibition of polycondensation reduced the mechanical strength (Figure 2). An increase in the NS/SS ratio promoted the dissolution of the initial solid, thus accelerating the geopolymerization reaction and leading to the rapid formation of geopolymer network structures. During the geopolymerization reaction, silicate derivatives were transformed into geopolymer gels. Therefore, the weight percentage of the silicate derivatives of 1.6NS0SCS decreased to 10.06% after 28 days of curing. Furthermore, the synergistic effect between the SCS and the MK was assigned to the promotion of the geopolymerization reaction, which caused the flexural strength of 1.6NS10SCS to increase to 6.42 MPa after 28 days of curing. First, the silica and alumina of metakaolin were dominant in the geopolymerization reaction, and the addition of silicon carbide sludge, which had silica, was provided more reaction paths. In addition, Rahman et al. [14] also noted that given the increase in flexural strength due to simultaneous inclusion of silica and silicon carbide whiskers, it is possible that silica particles are compatible with the metakaolin-based geopolymers, which is effective in consolidation [14]. According to the DTA/TG results, the weight loss percentage of 1.6NS10SCS increased to 14.62% from 400 to 750 °C, as shown in Figure 3b. This fact implies that the synergistic effect promoted the oligomers to combine and form integral geopolymer gels [20].

However, geopolymers with high replacement levels of SiC sludge generated an inadequate amount of leading precursors, thus limiting the polycondensation process. For example, for 1.6NS30SCS and 1.6NS40SCS, the weight percentages of silicate derivatives (Q^1, Q^2, and Q^3) were 9.81% and 9.13%, respectively, as presented in Table 4. In steps 1 and 2, SiC sludge dissolves to release Al^{3+} and Si^{4+}, which are hydrolyzed into [AlO_4] and [SiO_4] tetrahedrons, respectively, immediately after contact with the AASs. The amounts of dissolved silica and alumina from MK are limited due to precipitations of geopolymer gels around the surface of SCS particles, which caused the number of [SiO_4] and [AlO_4] tetrahedrons in the system to be low. Thus, the amount of oligomeric gel was low. Because the N-A-S-H gels activated by the sodium silicate solution required an increased amount of oligomeric gel, the rate of the geopolymerization reaction decreased. Therefore, the heat evolution duration of the second exothermic peak increased from 26.15 to 33.32 h as the replacement levels of SiC sludge increased from 30% to 40% (1.6NS30SCS and 1.6NS40SCS), as shown in Figure 1c. Additionally, the flexural strengths of 1.6NS30SCS and 1.6NS40SCS were 5.31 MPa and 2.73 MPa, respectively (Figure 2c). This result indicated that a relatively high SiC sludge content (more than 20%) might block the synergistic effect between SCS and MK, which is confirmed by the heat evolution rate, DTA/TG results, and ^{29}Si MAS NMR analyses. This is consistent with the SEM images results of previous studies [29] that the amorphous gel products had gradually filled the pores of the SiC sludge-based geopolymer, resulting in an increase in density and compressive strength of the geopolymer structure [29]. The synergistic effect of the SiC sludge and MK promoted the reaction progression, which caused increasing amounts of amorphously structured geopolymeric gels in the geopolymerized system [29]. Then, the unreacted SiC sludge was in the form of uniform plate particles [29]. In summary, this study demonstrated that this renders SiC sludge as promising additives for the production of metakaolin-based geopolymers. As a structural material, the SiC sludge-based geopolymer is a potential replacement material for OPC.

Table 4. ^{29}Si MAS NMR spectral deconvolution of the silicate derivatives (Curing Time = 1 day and 28 days).

Samples	Curing Time	Silicate Derivatives (%)			Deconvoluted Fractions of Silicon Centers (%)				
		Q^1	Q^2	Q^3	Q^4(0Al)	Q^4(1Al)	Q^4(2Al)	Q^4(3Al)	Q^4(4Al)
0.8NS0SCS	1 day	10.07	1.41	2.11	0.66	5.62	16.17	35.85	28.11
0.8NS10SCS		8.72	2.53	2.30	1.30	7.59	18.64	33.82	25.10
0.8NS20SCS		9.22	2.16	2.55	2.85	5.89	19.23	34.15	23.95
0.8NS30SCS		4.36	3.11	2.08	4.12	10.79	21.79	34.24	19.51
0.8NS40SCS		1.56	3.31	4.28	3.43	17.66	25.46	26.38	17.92
0.8NS0SCS	28 days	10.22	4.62	4.39	0.46	7.06	16.06	34.32	22.87
0.8NS10SCS		10.58	5.06	2.99	0.43	3.45	14.26	36.79	26.44
0.8NS20SCS		11.26	4.77	2.29	1.74	4.96	16.22	34.53	24.23
0.8NS30SCS		10.42	4.59	2.08	0.58	5.84	18.13	36.27	22.09
0.8NS40SCS		4.13	3.90	4.82	4.24	20.05	24.91	29.59	8.36
1.2NS0SCS	1 day	7.74	3.61	3.10	2.45	9.81	20.39	33.55	19.35
1.2NS10SCS		8.05	4.14	2.76	1.73	8.05	20.25	34.98	20.04
1.2NS20SCS		7.17	4.01	3.59	3.57	9.50	20.68	34.18	17.30
1.2NS30SCS		3.94	5.03	4.81	3.94	10.36	24.02	32.26	15.64
1.2NS40SCS		3.08	3.74	4.17	6.18	21.53	25.71	26.37	9.22
1.2NS0SCS	28 days	6.44	5.77	4.89	2.94	10.88	19.99	34.43	14.66
1.2NS10SCS		6.17	3.97	3.08	3.50	9.03	22.47	33.71	18.07
1.2NS20SCS		2.41	6.03	5.07	2.72	11.10	21.00	35.72	15.95
1.2NS30SCS		5.59	3.30	4.94	2.05	14.21	22.30	34.11	13.50
1.2NS40SCS		0.84	4.60	5.85	5.73	21.95	25.71	27.38	7.94
1.6NS0SCS	1 day	5.00	3.81	3.81	3.40	13.58	25.66	30.69	14.05
1.6NS10SCS		5.87	4.51	4.96	4.81	14.46	23.80	31.50	10.09
1.6NS20SCS		4.91	5.40	3.93	6.62	16.00	27.18	29.37	6.59
1.6NS30SCS		2.23	3.57	4.01	6.64	18.05	27.85	29.63	8.02
1.6NS40SCS		0.89	2.23	6.01	9.19	27.82	28.93	17.81	7.12
1.6NS0SCS	28 days	5.03	1.76	3.27	4.94	12.63	24.54	30.97	16.86
1.6NS10SCS		5.13	3.50	3.27	2.28	9.70	23.92	33.63	18.57
1.6NS20SCS		2.41	3.07	3.07	2.32	11.78	24.30	33.87	19.18
1.6NS30SCS		0.92	4.16	4.16	5.46	18.26	27.51	29.13	10.40
1.6NS40SCS		1.69	3.74	4.12	9.75	26.59	26.78	21.53	5.80

5. Conclusions

The present study investigated the effects of different SiC sludge contents and alkaline activator solutions on the geopolymer reaction and microstructures of metakaolin-based geopolymers subjected to different curing times. The geopolymer with the optimal SiC sludge replacement level and NS/SS ratio contained a high heat evolution value (990.6 W/g), excellent mechanical properties (6.42 MPa), and high deconvolution percentages of Q^4(3Al, 2Al, and 1Al). The conclusions can be summarized as follows. The high NaOH concentration prompted the dissolution of the initial solid in the alkaline activator, which caused the first exothermal peaks to be more intense. The 1.6NS10SCS (SiO_2/Al_2O_3 molar ratio of 1.51) had favorable mechanical strength due to an increasing SiO_2/Al_2O_3 molar ratio and a synergistic effect between the SiC sludge and the metakaolin. The geopolymers with high replacement levels of SiC sludge not only contained dissolved silica and alumina from metakaolin but also underwent polycondensation on a geopolymer reaction of SiC sludge-based geopolymers. A greater amount of SiC sludge (over 20%) may block the synergistic effect between the SiC sludge and metakaolin. The results of this study have shown that a geopolymer with an optimal SiC-sludge replacement level and NS/SS ratio expresses a synergistic effect that allows an improvement of both the geopolymer reaction and the strength of metakaolin-based geopolymers. This renders SiC sludge as promising additives for the production of metakaolin-based geopolymers. For future work, we expect to analyze other important properties of SiC sludge-based geopolymers by thermal conductivity, and fire performance, which was a promising alternative as fire resistance materials in the construction industry.

Author Contributions: Writing—original draft preparation, K.-W.L.; writing—review and editing, K.-L.L.; T.-W.C.; W.-T.L.; Y.-W.L. All authors have read and agreed to the published version of the manuscript.

Funding: This research was funded by Ministry of Science and Technology of the Republic of China, Taiwan (Contract No. MOST-109-2221-E-197-012-MY3).

Acknowledgments: The authors thank the Ministry of Science and Technology of the Republic of China, Taiwan, for supporting this research financially.

Conflicts of Interest: The authors declare no conflict of interest.

References

1. Chowdhury, M.S.; Rahman, K.S.; Chowdhury, T.; Nuthammachot, N.; Techato, K.; Akhtaruzzaman, M.; Tiong, S.K.; Sopian, K.; Amin, N. An overview of solar photovoltaic panels' end-of-life material recycling. *Energy Strategy Rev.* **2020**, *27*, 100431. [CrossRef]
2. Yoko, A.; Oshima, Y.S. Recovery of silicon from silicon sludge using supercritical water. *J. Supercrit. Fluids* **2013**, *75*, 1–5. [CrossRef]
3. Lan, A.; Liu, C.E.; Yan, H.L.; Yua, H.T.; Li, I.T.; Hsua, H.P.; Lan, C.W. Silicon ingot casting using reusable silicon nitride crucibles made from diamond wire sawing kerf-loss silicon. *J. Cryst. Growth* **2019**, *525*, 125184. [CrossRef]
4. Eblagon, F.; Ehrle, B.; Graule, T.; Kuebler, J. Development of siliconnitride/silicon carbide composites for wood-cutting tools. *J. Eur. Ceram. Soc.* **2007**, *27*, 419–428. [CrossRef]
5. Tsai, T.H. Silicon sawing waste treatment by electrophoresis and gravitational settling. *J. Hazard. Mater.* **2011**, *189*, 526–530. [CrossRef]
6. He, P.G.; Wang, M.R.; Fu, S.; Jia, D.C.; Yan, S.; Yuan, J.K.; Xu, J.H.; Wang, P.F.; Zhou, Y. Effects of Si/Al ratio on the structure and properties of metakaolin based geopolymer. *Ceram. Int.* **2016**, *42*, 14416–14422. [CrossRef]
7. Elyamany, H.R.; Elmoaty, A.E.M.A.; Elshaboury, A.M. Magnesium sulfate resistance of geopolymer mortar. *Constr. Build. Mater.* **2018**, *184*, 111–127. [CrossRef]
8. Leay, L.; Potts, A.; Donoclift, T. Geopolymers from fly ash and their gamma irradiation. *Mater. Lett.* **2018**, *227*, 240–242. [CrossRef]
9. Cui, Y.; Wang, D.; Wang, Y.; Sun, R.; Rui, Y. Effects of the n(H_2O: Na_2O_{eq}) ratio on the geopolymerization process and microstructures of fly ash-based geopolymers. *J. Non Cryst. Solids* **2019**, *511*, 19–28. [CrossRef]
10. Li, Z.; Zhang, S.; Zuo, Y.; Chen, W.; Ye, G. Chemical deformation of metakaolin based geopolymer. *Cem. Concr. Res.* **2019**, *120*, 108–118. [CrossRef]
11. Kolezyn'ski, A.; Kro'l, M.; Zychowicz, M. The structure of geopolymers-Theoretical studies. *J. Mol. Struct.* **2018**, *1163*, 465–471. [CrossRef]
12. Yusuf, M.O.; Johari, M.A.M.; Ahmad, Z.A.; Maslehuddin, M. Effects of H_2O/Na_2O molar ratio on the strength of alkaline activated ground blast furnace slag-ultrafine palm oil fuel ash based concrete. *Mater. Des.* **2014**, *56*, 158–164. [CrossRef]
13. Medri, V.; Ruffini, A. Alkali-bonded SiC based foams. *J. Eur. Ceram. Soc.* **2012**, *32*, 1907–1913. [CrossRef]
14. Rahman, A.S.; Shah, C.; Gupta, N. Simultaneous effects of rice husk silica and silicon carbide whiskers on the mechanical properties and morphology of sodium geopolymer. *J. Compos. Mater.* **2020**, 0021998320928579. [CrossRef]
15. Medri, V.; Fabbri, S.; Ruffini, A.; Dedecek, J.; Vaccari, A. SiC-based refractory paints prepared with alkali aluminosilicate binders. *J. Eur. Ceram. Soc.* **2011**, *31*, 2155–2165. [CrossRef]
16. Bai, C.Y.; Zheng, J.; Rizzi, G.A.; Colombo, P. Low-temperature fabrication of SiC/geopolymer cellular composites. *Compos. Part B* **2018**, *137*, 23–30. [CrossRef]
17. Jia, D.C.; Li, Y.H.; He, P.G.; Fu, S.; Duan, X.M.; Sun, Z.L.; Cai, D.; Li, D.X.; Yang, Z.H.; Zhou, Y. In-situ formation of bulk and porous h-AlN/SiC-based ceramics from geopolymer technique. *Ceram. Int.* **2019**, *45*, 24727–24733. [CrossRef]

18. Wan, Q.; Rao, F.; Song, S.X. Reexamining calcination of kaolinite for the synthesis of metakaolin geopolymers-roles of dehydroxylation and recrystallization. *J. Non Cryst. Solids.* **2017**, *460*, 74–80. [CrossRef]
19. Wei, B.; Zhang, Y.; Bao, S. Preparation of geopolymers from vanadium tailings by mechanical activation. *Constr. Build. Mater.* **2017**, *145*, 236–242. [CrossRef]
20. Hu, Y.; Liang, S.; Yang, J.K.; Chen, Y.; Ye, N.; Ke, Y.; Tao, S.Y.; Xiao, K.K.; Hu, J.P.; Hou, H.J.; et al. Role of Fe species in geopolymer synthesized from alkali-thermal pretreated Fe-rich Bayer red mud. *Constr. Build. Mater.* **2019**, *200*, 398–407. [CrossRef]
21. Lo, K.W.; Lin, K.L.; Cheng, T.W.; Shiu, H.S. Effect of alkali activation thin film transistor-liquid crystal display waste glass on the mechanical behavior of geopolymers. *Constr. Build. Mater.* **2018**, *162*, 724–731. [CrossRef]
22. Yaghoubi, M.; Arulrajah, A.; Disfani, M.M.; Horpibulsuk, S.; Darmawan, S.; Wang, J. Impact of field conditions on the strength development of a geopolymer stabilized marine clay. *Appl. Clay Sci.* **2019**, *167*, 33–42. [CrossRef]
23. Henon, J.; Alzina, A.; Absi, J.; Smith, D.S.; Rossignol, S. Porosity control of cold consolidated geomaterial foam: Temperature effect. *Ceram. Int.* **2012**, *38*, 77–84. [CrossRef]
24. Tiffo, E.; Mbah, J.B.B.; Belibi, P.D.B.; Djobo, J.N.Y.; Elimbi, A. Physical and mechanical properties of unheated and heated kaolin based-geopolymers with partial replacement of aluminium hydroxide. *Mater. Chem. Phys.* **2020**, *239*, 122103. [CrossRef]
25. Wan, Q.; Rao, F.; Song, S.X.; León–Patiño, C.A. Geothermal clay-based geopolymer binders: Synthesis and microstructural characterization. *Appl. Clay Sci.* **2017**, *146*, 223–229. [CrossRef]
26. Li, W.; Rao, F.; Song, S.X.; Ma, Q.Y. Deterioration in the microstructure of metakaolin-based geopolymers in marine environment. *J. Mater. Res. Technol.* **2019**, *8*, 2747–2752. [CrossRef]
27. Maekawa, H.; Maekawa, T.; Kawamura, K.; Yokokawa, T. The structural groups of alkali silicate glasses determined from ^{29}Si Mas-NMR. *J. Non Cryst. Solid.* **1991**, *127*, 53–64. [CrossRef]
28. Wan, Q.; Rao, F.; Song, S.X.; García, R.E.; Estrella, R.M.; Patino, C.L.; Zhang, Y.M. Geopolymerization reaction, microstructure and simulation of metakaolin-based geopolymers at extended Si/Al ratios. *Cem. Concr. Compos.* **2017**, *79*, 45–52. [CrossRef]
29. Lin, K.L.; Lo, K.W.; Cheng, T.W.; Lin, W.T.; Lin, Y.W. Influence of SiC Sludge on the Microstructure of Geopolymers. *Materials* **2020**, *13*, 2203. [CrossRef]

sustainability

MDPI

Article

Electromagnetic Shielding Effectiveness and Conductivity of PTFE/Ag/MWCNT Conductive Fabrics Using the Screen Printing Method

Hung-Chuan Cheng [1], Chong-Rong Chen [2], Shan-hui Hsu [1,*] and Kuo-Bing Cheng [2,3,*]

1 Institute of Polymer Science and Engineering, National Taiwan University, Taipei 106, Taiwan; d02549012@ntu.edu.tw
2 Department of Fiber and Composite Material, Feng Chia University, Taichung 407, Taiwan; zx25610548as@gmail.com
3 Textile and Material Industrial Research Center, Feng Chia University, Taichung 407, Taiwan
* Correspondence: shhsu@ntu.edu.tw (S.-h.H.); kbcheng@fcu.edu.tw (K.-B.C.)

Received: 23 June 2020; Accepted: 20 July 2020; Published: 22 July 2020

Abstract: The management of the electromagnetic interference (EMI) of thin, light, and inexpensive materials is important for consumer electronics and human health. This paper describes the development of conductive films that contain a silver (Ag) flake powder and multiwall carbon nanotube (MWCNT) hybrid grid on a polytetrafluoroethylene (PTFE) film for applications that require electromagnetic shielding (EMS) and a conductive film. The Ag and MWCNT hybrid grid was constructed with a wire diameter and spacing of 0.5 mm. The results indicated that the proposed conductive films with 0.4 wt% MWCNTs had higher electromagnetic shielding effectiveness (EMSE) and electrical conductivity than those with other MWCNT loading amounts. The results also showed that the film with 0.4 wt% MWCNT loading had a high 62.4 dB EMSE in the 1800 MHz frequency and 1.81×10^4 S/cm electrical conductivity. This combination improved stretchability, with 10% elongation at a 29% resistivity change rate. Conductive films with Ag/MWCNT electronic printing or lamination technologies could be used for EMI shielding and electrically conductive applications.

Keywords: Ag/MWCNT; EMSE; laminated woven fabric; PTFE film; screen printing

1. Introduction

Traditional materials that can be used for electromagnetic shielding are metal, ceramic, and plastic materials. During low-frequency magnetic field shielding, absorption loss is more important than reflection loss, so steel, plastic materials, or high-permeability magnetic materials are generally used; during high-frequency magnetic field shielding, reflection loss is rather critical, thus, aluminum foil or conductive coating materials are selected [1]. With the advent of the 5th generation (5G) mobile network, self-driving cars, cloud computing, and various electronic products are accelerating the promotion of various specifications of the wireless network. The number of small-cell base stations for the 5G millimeter-wave is expected to be 60 times that of traditional 4G stations. In addition, the large number of sensing elements used in self-driving systems means that people will be exposed to GHz UHF (ultrahigh frequency) transmission; thus, the issue of the impact of electromagnetic waves on human health will come to the fore [2,3]. Additionally, the high transmission efficacy and large number of high-power components that are used in radio frequency (RF) antenna modules and base station basic circuits, e.g., integrated circuits (ICs), system on a chips (SoCs), system in packages (SiPs), antenna in packages (AiPs), power amplifiers, switches, and passive components, cause substantial heat problems. The construction of a healthy indoor environment in modern buildings that is comfortable, convenient, and safe, for example, is the topic of discussion herein. In general, as long as the electrical equipment is properly grounded and the transformer equipment is shielded by electromagnetic shields in new residential projects, the harm from low-frequency electric fields can be greatly reduced. However, old houses are not grounded, and the source of the electromagnetic waves is outdoors or cannot be solved with high-frequency and ultrahigh frequency electromagnetic radiation; thus, this situation presents a risk to human health.

In the application of EMI shielding films, the conductivity and flexibility of conductive materials are directly related to EMSE and reliability. The migration process of electrons is a key factor affecting the overall electrical properties of conductive composites. There are generally three well-known mechanisms of electrical conductivity in composite materials: percolation, tunneling, and field emission. Percolation theory describes the conductive behavior when the proportion of conductive filler reaches a certain content, and the filler materials are in close contact or the gap is less than 1 nm. The tunneling effect occurs when the applied voltage is low and the distance between the filler materials is less than 10 nm; even if there is no direct contact between the filler materials, electrons can still pass through the gap to achieve conduction. The field emission effect occurs when the filler content is low, but the application of a high electric field can cause the migration of electrons through the isolation layer [4–6].

Based on the price factor and the component requirement, silver paste is the most used in the application of traditional thick film products (e.g., printed circuit boards, membrane switches, passive component electrodes, photovoltaics, radio frequency identification (RFID) devices, and'sensors). The addition of multiwall carbon nanotubes (MWCNTs) can increase the probability of conductive fillers contacting each other and can prevent conductivity from decreasing when the distance between the conductive fillers is greater than 10 nm. In addition, a small amount of MWCNTs can increase electrical conductivity, thermal conductivity, and mechanical strength [7–10]. The addition of MWCNTs has a cost advantage over simply adding conductive fillers.

This research provides a method to improve high-frequency electromagnetic shielding and thermal properties [11]. By introducing MWCNTs to increase the conductivity and stretchability of conductive silver paste, screen printing can be used to print an effective shielding circuit [12–15]. A polytetrafluoroethylene (PTFE) substrate can be attached to the nonwoven EMI shielding film with a low manufacturing cost, high EMSE, and high reliability [16–19]. Compared with the weight and thickness of steel, ceramic materials, and plastic materials, the samples made in this experiment can be easily applied to curtains, microwave ovens, or electronic equipment to cover woven fabrics, substation shielding films, smart textile wearable products, and protective clothing, for example.

2. Experimental Details

2.1. Preparation of Ag/MWCNT Hybrid Conductive Paste

A well-dispersion of 4 wt% MWCNT was prepared by combining a specific weight of gum arabic (GA, 0.14 g), poly(acrylic acid sodium salt) (PAAS, 0.06 g), isopropyl alcohol (IPA, 9.6 g), and MWCNTs (0.4 g) in ultrasonic and planetary centrifugal mixers (Figure 1) [20,21].

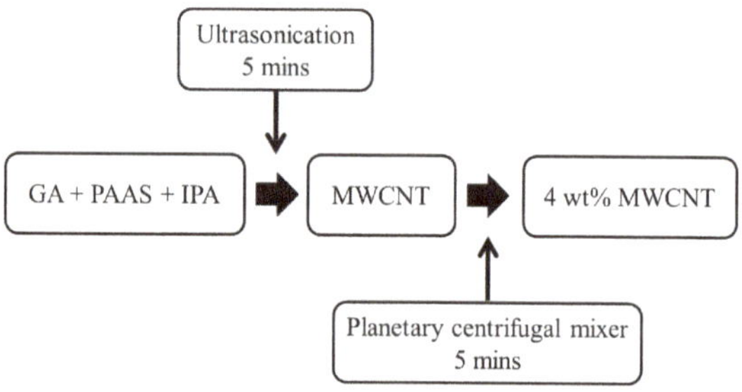

Figure 1. The preparation process of 4 wt% multiwall carbon nanotube (MWCNT) dispersion.

Ten grams of conductive Ag/MWCNT composite paste were used for the preparation and formulation processes shown in Figure 2 and Table 1. Appropriate amounts of silver paste and 4 wt% MWCNT dispersion, with weights (g) of 9.98:0.35, 9.97:0.5, 9.96:0.68, 9.95:0.85, and 9.94:1.03, respectively, were formulated. The concentrations of MWCNT of 0.2, 0.3, 0.4, 0.5, and 0.6 wt% in the Ag/MWCNT composite paste were prepared by a planetary centrifugal mixer for 2 minutes. The referenced codes refer to the different MWCNT contents, such as 4SC indicating a solid content of MWCNT (SCM) that accounts for 0.4 wt% of the solid content of silver paste (SCSP).

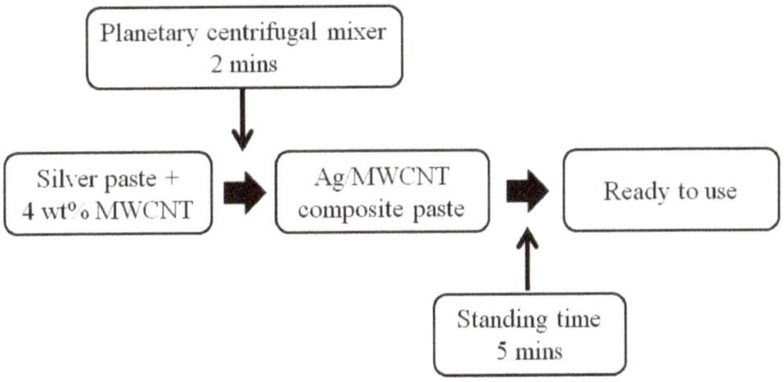

Figure 2. The preparation process of Ag/MWCNT pastes.

Table 1. The compositions of Ag/MWCNT hybrid conductive pastes.

	Silver Paste (g)	***1 SCSP (wt%)**	**MWCNT Dispersion (g)**	***2 SCM (wt%)**	**SCM/SCSP (wt%)**
Reference	10	6.8	-	-	-
2SC	9.98	6.786	0.35	0.014	0.2
3SC	9.97	6.78	0.5	0.02	0.3
4SC	9.96	6.773	0.68	0.027	0.4
5SC	9.95	6.766	0.85	0.034	0.5
6SC	9.94	6.759	1.03	0.041	0.6

*1 Solid Content of Silver Paste (SCSP). *2 Solid Content of MWCNT (SCM).

The quality of raw materials is very important when we do formula modification. It needs to balance the compatibility, electrical properties, and rheology for screen printing of the materials. Thus, measurements of 2.5 to 20 um for the particle size of the silver flake, 58±2% for silver solid content, and 23,000 cps for viscosity can approach the highest electrical conductivity and good bending ability of the conductive silver paste (AG-1003, HongChen Materials Corporation, Taiwan). The average diameter of the MWCNTs was 8 to 15 nm, the length was 10 to 70 µm, the density was 0.1 to 0.15 g/cm^3, and the surface area was 235 to 275 m^2/g (CP1002M, LG Chemical, Korea) (Figure 3).

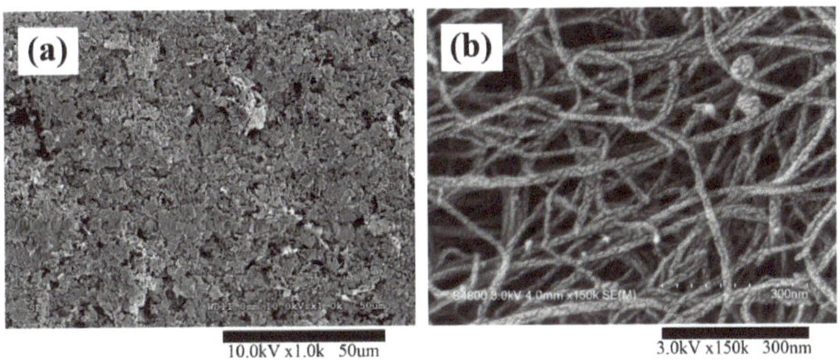

Figure 3. Variable vacuum scanning electron microscopy (VVSEM) image of the (**a**) conductive silver paste and (**b**) MWCNT.

2.2. Production of Conductive Laminated Fabrics

A screen printing method was used to fabricate conductive laminated fabrics for this paper. A viscous and conductive composite paste that contained a suitable amount of silver flakes and MWCNTs, thermal plastic urethanes, a multi-solvent system, and additive agents was used during fabrication. The Ag/MWCNT composite paste was fed onto a 300-mesh screen and manipulated with a squeegee to form a grid pattern, in order to save on the material's cost. The grid pattern was constructed with a wire diameter and spacing of 0.5 mm. The recovery area was about 75% of the PTFE film, and thus provided a cost-efficient solution. Adding a certain amount of MWCNTs and making a composite paste with conductive silver paste provided the substantial benefits of reducing raw material cost and improving conductivity.

The PTFE/Ag/MWCNT composite film was laminated to the cotton woven fabric, where polyurethane reactive glue (PUR Glue) was the adhesive layer, with 120 ends/2.54 cm and 80 picks/2.54 cm warp and weft densities, respectively (Figure 4). The warp and weft yarn count was 59.1 tex. The tested specimens used in this investigation are shown in Figure 4. The specimens had a nominal width and length of 15 cm. An Ag/MWCNT conductive circuit was shown on the PTFE substrate, then the printed side of the conductive circuit was laminated onto the woven fabric with PUR glue, then the release paper was removed.

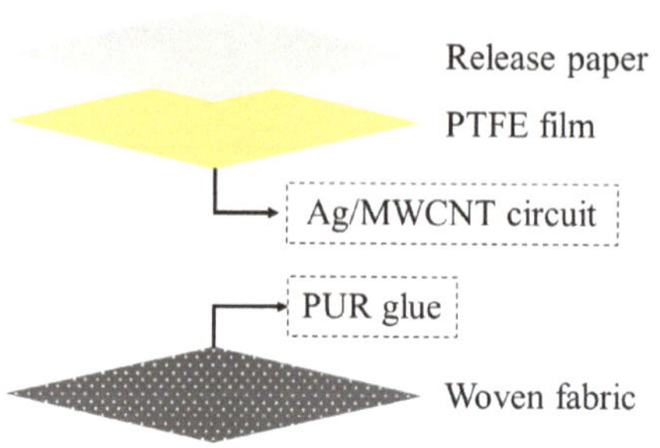

Figure 4. The structure of the laminated fabric (size: 15*15 cm).

2.3. Surface Resistivity, Volume Resistivity and Electrical Conductivity Measurements

Measurements were made using a Loresta-GP MCP-T600 resistivity meter (Mitsubishi Chemical Corporation, Japan), followed by a JIS K 7194 [22]. In order to confirm the stability of the Ag/MWCNT composite paste and screen printing process, volume resistivity (ρ_v) was checked for 3 to 10 samples, and the results show a window within 5% [23].

The stretchability testing was conducted according to ASTM D638 [24], and the proposed bending testing methods reproduced the state of cyclic unloading conditions for 180 degrees by placing the samples between two parallel plates to observe the resistance of the coefficient of variation [25].

2.4. Morphology of PTFE/Ag/MWCNT Conductive Composite Film

The morphology of the conductive circuit was observed using high-resolution variable vacuum scanning electron microscopy (HR-VVSEM).

2.5. Measurement of the Laminated Fabric of PTFE/Ag/MWCNT Conductive Composite Film

The EMSE of the PTFE/Ag/MWCNT conductive composite films was measured by an Elgal set 19A coaxial holder system (EM-21078, ShangHai YinXu Mechanical and Electrical Equipment Co., Ltd., China) [26,27]. The thermal conductivity (mW/m.K) and thermal diffusivity (mm^2/S) were measured by a thermal property analyzer (Alambeta, Sensor Instruments & Consulting REG.

NR., Czech Republic) [28]. An infrared thermal imaging camera (FLIR-A320, FLIR Systems Inc., USA) was used to observe the temperature difference of the PTFE/Ag/MWCNT conductive composite film [29].

3. Results and Discussion

3.1. Morphological Analysis of PTFE/Ag/MWCNT Composite Film

In this study, the quality of the Ag/MWCNT composite paste and the morphology of the circuit are very important due to the correlation of conductivity, skin effect, and EMSE. The electric current flows mainly along the skin of the conductor and does not penetrate to the core. The skin depth decreases as the radio frequency increases, so the surface roughness of the circuit is one of the major parameters that needs to be checked when high incident frequency applications are used. The SEM images of the Ag/MWCNT conductive circuit in sample 4SC are shown in Figure 5a. However, it is difficult to produce hand-made samples with high uniformity (Figure 5b), resulting in defects in the circuit and poor EMSE. Figure 5c,d show that MWCNTs can provide a complete conductive network to improve electrical conductivity.

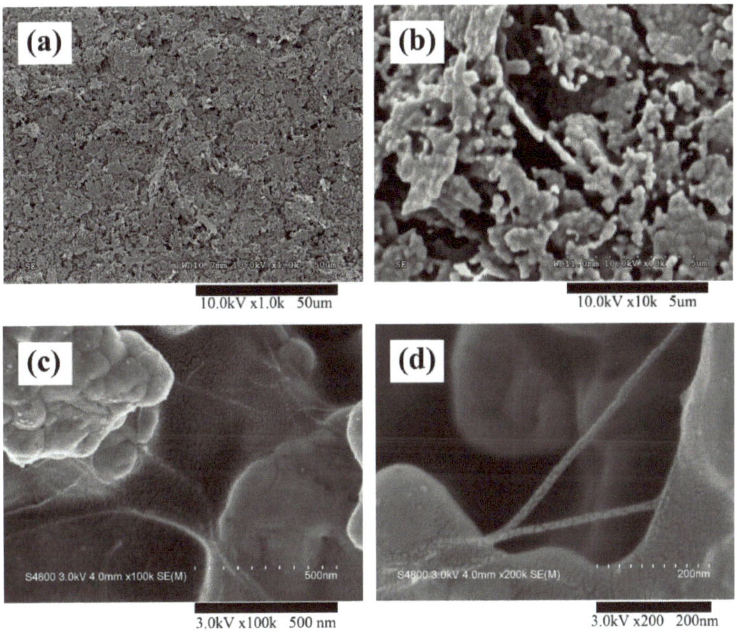

Figure 5. VVSEM images of the Ag/MWCNT conductive circuit in sample 4SC. (**a**) Topography, (**b**) morphology and arrangement of the silver flakes, (**c**,**d**) connections between the MWCNTs and silver flakes.

A conductive PTFE/Ag/MWCNT composite film was produced with a thermal curable Ag/MWCNT composite conductive paste and a PTFE substrate. A good connection of the conductivity of Ag/MWCNT fillers, a specified grid pattern, and a PTFE substrate improved the electrical properties, flexibility, and feel of the fabric, and saved on the material's consumption. Since the surface roughness of PTFE substrate is smaller than fabric, it can ameliorate the grid pattern of the printed situation to get better and more stable electrical conductivity, flexibility, and EMSE. Through changing the recipe of the Ag/MWCNT composite paste, the rheology was changed. The optical microscope (OM) image of Figure 6b (4SC) shows better printing quality than Figure 6a (pure silver). This result demonstrates that higher viscosity was more suitable for the hand-made screen printing process.

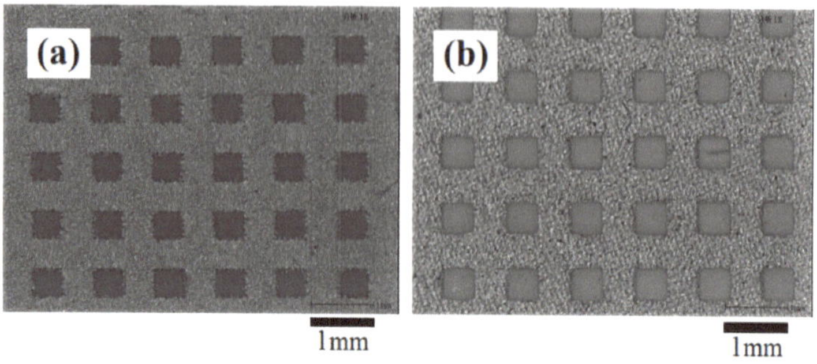

Figure 6. Optical microscope (OM) image of the grid pattern of (**a**) pure conductive silver paste and (**b**) 4SC.

3.2. Electrical and Mechanical Properties of PTFE/Ag/MWCNT Composite Film

Figure 7a shows the variation in the conductivity of the PTFE/Ag/MWCNT composite films when changing the amount of MWCNTs added to the conductive Ag/MWCNT composite paste. To produce much smoother mesh wires, dibasic esters (DBE) can be a thinner to adjust the rheology of the Ag/MWCNT conductive paste, which makes the process of screen printing easier, and optimal electrical conductivity was obtained with 0.4 wt% MWCNT loading (4SC). For the composite film, electrical conductivity reached $1.81*10^4$ S/cm, with the 6SC film having the lowest conductivity. The maximum conductivity of the 4SC composite film is 30% higher than the conductivity of the 6SC film. The conductivity trends of PTFE/Ag/MWCNT films demonstrated that loading MWCNTs did not do much good. This can be explained by percolation theory. MWCNT loading had a notable influence on the electrical properties of the films. Thus, conductivity can be modified through different component ratios of silver powder and MWCNTs (by varying the concentration, uniformity of the conductive network, or Ag/MWCNT wire density). Figure 7b shows the results in the surface resistivity of the PTFE/Ag/MWCNT conductive films as a function of the amount of MWCNTs added to the conductive Ag/MWCNT composite paste. The optimal surface resistivity also occurred with a loading of 0.4 wt% MWCNT (4SC), whereas the 6SC film had the highest surface resistivity.

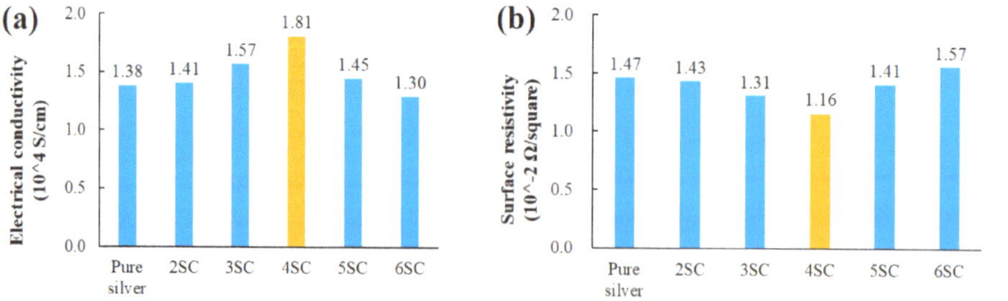

Figure 7. Variation in the (**a**) conductivity (Siemens/cm) and (**b**) surface resistivity of the conductive PTFE/Ag/MWCNT films.

When electronic products, such as flexible substrates, are used, the products undergo a certain degree of bending and deformation, causing circuit damage that impacts electrical conductivity. In textile products, it is necessary to consider the damage caused by multiple cleanings and laundry. At present, an increasing number of substrates are malleable. Therefore, conductive materials that can withstand tensile and bending resistance are becoming important. The effects of tensile elongation and hybrid paste formulation on the changing percentage of electrical resistance of the PTFE/Ag/MWCNT conductive films are shown in Figure 8a. The PTFE/Ag/MWCNT conductive films had an improved variation in electrical resistance under 10% tensile elongation. The improvement in flexibility can be seen in the stretchability and cyclic bending test results. Through the coefficient of variation, it can be observed that under 10% elongation conditions, 3SC (16.7%) and 4SC (15.1%) showed increased stability (Figure 8b). Similar results can be seen in the cyclic bending test (0% for 4SC and 5SC) (Figure 9). It can be seen from the cyclic bending test that there was an increase in the number of cyclic bending cycles of the composite film. The change in resistivity was extremely small because MWCNTs are very strong, their mechanical strength is very good, and it is not easy to break the composite film after cyclic bending. Under 1000 cycles of the bending test, the coefficient of variation did not change for 4SC and 5SC, which showed excellent reproducibility, reliability, and stability (Figure 9). The high-aspect-ratio MWCNTs contributed to the multiple conductive pathways, tough mechanical properties, flexibility, and strength of the Ag/MWCNT composite paste under testing.

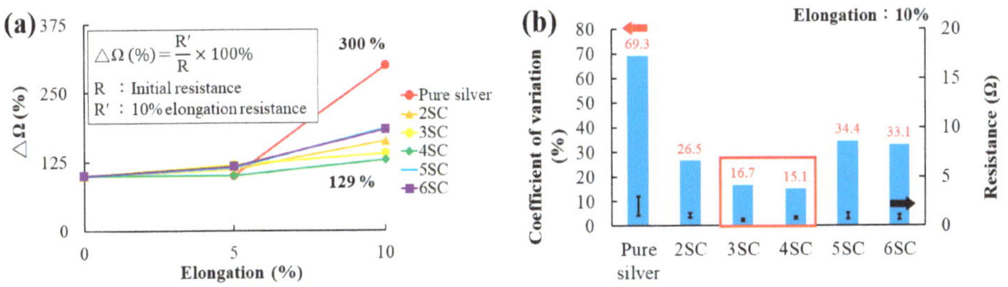

Figure 8. (**a**) The effects of tensile elongation and hybrid paste formulation on the changing percentage of the electrical resistance of PTFE/Ag/MWCNT conductive films and (**b**) the coefficient variance of the electrical resistance of PTFE/Ag/MWCNT film under 10% tensile elongation.

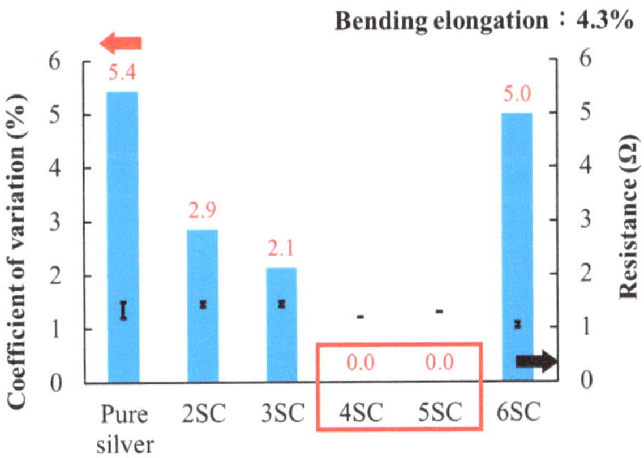

Figure 9. CV% of the electrical resistance of the PTFE/Ag/MWCNT conductive films under 1000 cycles of bending testing.

3.3. EMSE Analysis of PTFE/Ag/MWCNT Conductive Laminated Fabrics

The EMSE results (measurement frequency is 5 MHz to 3000 MHz) of the conductive laminated fabrics are presented in Figure 10 and Table 2. A regular up and down pattern with frequencies from 5 MHz to 3000 MHz is shown. The PTFE/Ag/MWCNT conductive films with an addition of 0.4 wt% MWCNTs displayed the highest EMSE, compared to the other PTFE/Ag/MWCNT films. According to the electromagnetic wave shielding test, the PTFE/Ag/MWCNT conductive laminated fabric printed with the Ag/MWCNT composite paste had a good electromagnetic wave shielding effect, and its EMSE value exceeded 40 dB. At a frequency of 300 MHz (domestic wireless telephone), the 3SC has a good EMSE value of 50.1 dB; at a frequency of 1800 MHz (4G mobile phone), 4SC has the best EMSE value of 62.4 dB; at a frequency of 2450 MHz (microwave oven), 4SC has the best EMSE of 59.9 dB. The results reach very good to excellent levels, according to the classifications of EMSE for general purposes (Table 3).

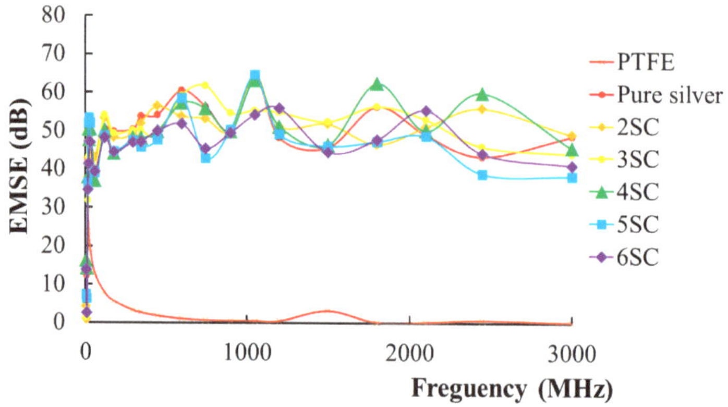

Figure 10. EMSE vs. frequency plots for different MWCNT contents in the PTFE/Ag/MWCNT conductive laminated fabrics.

Table 2. EMSE data for the applications of 300 MHz, 1800 MHz, and 2450 MHz.

Application/Unit: dB	PTFE	Pure silver	2SC	3SC	4SC	5SC	6SC
300 MHz (Cordless telephone)	3.1	50.3	48.9	50.1	48.0	47.5	47.1
1800 MHz (4G mobile phone)	0.1	56.3	46.5	56.4	62.4	47.3	47.8
2450 MHz (Microwave oven)	0.6	43.3	55.9	46	59.9	38.8	44.1

Table 3. The classifications of EMSE for general purposes.

Type	Class 5 Excellent	Class 4 Very Good	Class 3 Good	Class 2 Moderate	Class 2 Fair
EMSE ranges	SE > 60	60 ≥ SE > 50	50 ≥ SE > 40	40 ≥ SE > 30	30 ≥ SE > 20

3.4. Thermal Analysis of PTFE/Ag/MWCNT Conductive Laminated Fabrics

Table 4 shows the thermal properties of PTFE/Ag/MWCNT conductive laminated fabric. It can be seen from the thermal property tests that with an increase in the amount of MWCNTs, the thermal conductivity of the laminated fabric tended to rise first and then decrease because the addition of the MWCNTs correspondingly reduced the silver solid content, resulting in a downward trend. Sample 4SC showed the highest thermal conductivity coefficient and thermal diffusion values of 37.1 mW/m.K and 0.240 mm^2/s, respectively. This is probably because the 4SC tested sample had the best thermal conductive network. However, as the MWCNT amount increased, agglomeration increased.

Figure 11 and Table 5 present the results in the temperature difference of the conductive laminated fabrics for the different formulations. The temperature difference of the laminated fabric increased first and then decreased because the addition of MWCNTs correspondingly reduced the silver solid content, resulting in a downward trend. From the far infrared thermal imaging test, it can be seen that the laminated fabric with the added MWCNTs had a better heating effect than the fabric that did not contain the MWCNTs because the MWCNTs have high thermal conductivity and good heat transfer characteristics. The 4SC has the highest temperature increase of 5.6 °C and temperature decrease of 4.0 °C.

Table 4. Thermal properties of PTFE/Ag/MWCNT conductive laminated fabrics.

Sample/Data	Pure Silver	2SC	3SC	4SC	5SC	6SC
Thermal conductivity (mW/m.K)	35.5	36.3	36.7	37.1	36.5	36.2
Thermal diffusivity (mm^2/s)	0.132	0.157	0.217	0.240	0.200	0.118

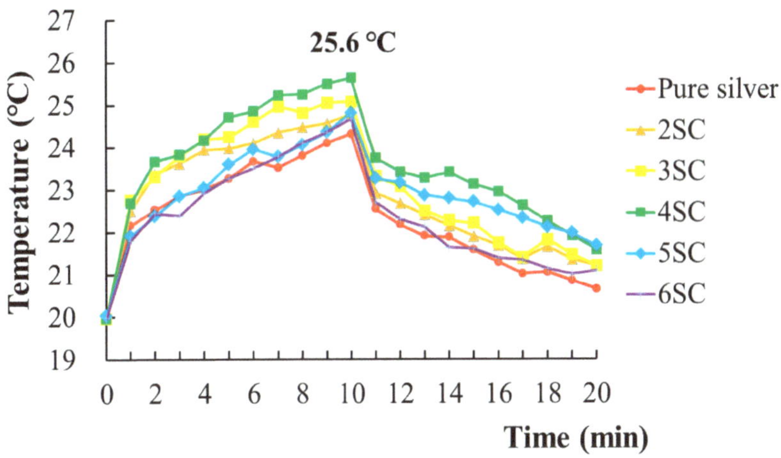

Figure 11. Temperature changes under far infrared irradiation of PTFE/Ag/MWCNT conductive laminated fabrics.

Table 5. Temperature changes during far-infrared irradiation of PTFE/Ag/MWCNT conductive laminated fabrics.

Sample	T0	T10	T20	ΔT1 (T10-T0)	ΔT2 (T10-T20)	ΔT3 (T20-T0)
Pure silver	20.0	24.3	20.7	4.3	3.6	0.7
2SC	20.0	24.8	21.2	4.8	3.6	1.2
3SC	20.0	25.1	21.2	5.1	3.9	1.2
4SC	20.0	25.6	21.6	5.6	4.0	1.6
5SC	20.0	24.8	21.7	4.8	3.1	1.7
6SC	20.0	24.7	21.1	4.7	3.6	1.1

4. Conclusions

In this paper, an Ag/MWCNT composite conductive paste was applied to a PTFE film, using a screen printing method to produce a grid pattern with high electrical conductivity, EMSE, thermal conductivity, and diffusivity functions. This PTFE/Ag/MWCNT composite film was then used to laminate woven fabric, providing a low cost material with highly stable electrical properties for printing, high durability, good softness, high EMSE, and improved thermal properties. The EMSE of the PTFE/Ag/MWCNT conductive laminated woven fabrics can be adjusted via designing different grid patterns and controlling the ratio of Ag/MWCNT. It can be observed that PTFE/Ag/MWCNT conductive laminated fabrics may have applications as home textiles which solve health and electrical issues related to electromagnetic shielding and thermal problems, being especially easy to use because of their flexibility. The 4SC conductive fabric sample had a higher EMSE than other specimens for almost all frequency ranges from 120 to 3000 MHz. The EMSE and thermal properties of the PTFE/Ag/MWCNT laminated fabrics showed improved performance and are suitable for use in domestic electronic appliances and household textiles. Since MWCNT shows excellent improvement of flexibility and elongation, it has the potential to acquire the stretchable property of conductive ink in the future.

Author Contributions: Conceptualization, H.-C.C.; methodology, H.-C.C.; validation, H.-C.C. and C.-R.C.; formal analysis, C.-R.C.; investigation, C.-R.C.; resources, H.-C.C. and K.-B.C.; data curation, C.-R.C.; writing—original draft preparation, H.-C.C.; writing—review and editing, S.-h.H. and K.-B.C.; visualization, C.-R.C. and H.-C.C.; supervision, S.-h.H. and K.-B.C.; project administration, K.-B.C.; funding acquisition, K.-B.C. All authors have read and agreed to the published version of the manuscript.

Funding: This research was funded by the Ministry of Science and Technology (MOST) of Taiwan to Feng Chia University under MOST-108-2218-E-035-001.

Acknowledgments: This work was supported by the Ministry of Science and Technology (MOST) of Taiwan to Feng Chia University under MOST-108-2218-E-035-001. The authors also thank the Institute of Polymer Science and Engineering, National Taiwan University, Taiwan, for guidance on materials and formulation principles, and the Fiber Structure Lab of Feng Chia University, Taiwan, for providing all necessary instruments.

Conflicts of Interest: The authors declare no conflicts of interest.

References

1. Shukla, V. Review of electromagnetic interference shielding materials fabricated by iron ingredients. *Nanoscale Adv.* **2019**, *1*, 1640–1671. [CrossRef]
2. Al-Falahy, N.; Alani, O.Y. Technologies for 5G Networks: Challenges and Opportunities. *IT Prof.* **2017**, *19*, 12–20. [CrossRef]
3. Zhao, K.; Ying, Z.; He, S. EMF Exposure Study Concerning mmWave Phased Array in Mobile Devices for 5G Communication. *IEEE Antennas Wirel. Propag. Lett.* **2016**, *15*, 1132–1135. [CrossRef]
4. He, L.-X.; Tjong, S.-C. Internal field emission and conductivity relaxation in carbon nanofiber filled polymer system. *Synth. Met.* **2010**, *160*, 2085–2088. [CrossRef]
5. Hashemi, R.; Weng, G.J. A theoretical treatment of graphene nanocomposites with percolation threshold, tunneling-assisted conductivity and microcapacitor effect in AC and DC electrical settings. *Carbon* **2016**, *96*, 474–490. [CrossRef]
6. Bilotti, E.; Zhang, H.; Deng, H.; Zhang, R.; Fu, Q.; Peijs, T. Controlling the dynamic percolation of carbon nanotube based conductive polymer composites by addition of secondary nanofillers: The effect on electrical conductivity and tuneable sensing behavior. *Compos. Sci. Technol.* **2013**, *74*, 85–90. [CrossRef]
7. Chun, K.-Y.; Oh, Y.; Rho, J.; Ahn, J.-H.; Kim, Y.-J.; Choi, H.R.; Naik, S. Highly conductive, printable and stretchable composite films of carbon nanotubes and silver. *Nat. Nanotechnol.* **2010**, *28*, 853–857. [CrossRef]
8. Maiti, S.; Suin, S.; Shrivastava, N.K.; Khatua, B.B. A strategy to achieve high electromagnetic interference shielding and ultra low percolation in multiwall carbon nanotube-polycarbonate composites through selective localization of carbon nanotubes. *RSC Adv.* **2014**, *4*, 7979–7990. [CrossRef]
9. Rosca, I.D.; Hoa, S.V. Highly conductive multiwall carbon nanotube and epoxy composites produced by three-roll milling. *Carbon* **2009**, *47*, 1958–1968. [CrossRef]
10. Ma, R.; Kwon, S.; Zheng, Q.; Kwon, H.Y.; Kim, J.I.; Choi, H.R.; Baik, S. Carbon-Nanotube/Silver Networks in Nitrile Butadiene Rubber for Highly Conductive Flexible Adhesives. *Adv. Mater.* **2012**, *24*, 3344–3349. [CrossRef]
11. Pawar, S.P.; Biswas, S.; Kar, G.P.; Bose, S. High frequency millimetre wave absorbers derived from polymeric nanocomposites. *Polymer* **2016**, *84*, 398–419. [CrossRef]
12. Hassan, M.; Afify, A.S.; Ataalla, M.; Milanese, D.; Tulliani, J.-M. New ZnO-Based Glass Ceramic Sensor for H_2 and NO_2 Detection. *Sensors* **2017**, *17*, 2538. [CrossRef] [PubMed]
13. Afify, A.S.; Ahmad, S.; Khushnood, R.A.; Jagdale, P.; Tulliani, J.-M. Elaboration and characterization of novel humidity sensor based on micro-carbonized bamboo particles. *Sens. Actuators B Chem.* **2017**, *239*, 1251–1256. [CrossRef]

14. Erath, D.; Filipović, A.; Retzlaff, M.; Goetz, A.K.; Clement, F.; Biro, D.; Preu, R. Advanced screen printing technique for high definition front side metallization of crystalline silicon solar cells. *Sol. Energy Mater. Sol. Cells* **2010**, *94*, 57–61. [CrossRef]
15. Hsu, C.-P.; Guo, R.-H.; Hua, C.-C.; Shih, C.-L.; Chen, W.-T.; Chang, T.-I. Effect of polymer binders in screen printing technique of silver pastes. *J. Polym. Res.* **2013**, *20*, 277. [CrossRef]
16. Kolanowska, A.; Janas, D.; Herman, A.P.; Jędrysiak, R.G.; Giżewski, T.; Boncel, S. From blackness to invisibility—Carbon nanotubes role in the attenuation of and shielding from radio waves for stealth technology. *Carbon* **2018**, *126*, 31–52. [CrossRef]
17. Kim, Y.J.; An, K.J.; Suh, K.S.; Choi, H.-D.; Kwon, J.H.; Chung, Y.-C.; Kim, W.N.; Lee, A.-K.; Choi, J.-I.; Yoon, H.G. Hybridization of oxidized MWNT and silver powder in polyurethane matrix for electromagnetic interference shielding application. *IEEE Trans. Electromagn. Compat.* **2005**, *47*, 872–879. [CrossRef]
18. Shajari, S.; Arjmand, M.; Pawar, S.P.; Sundararaj, U.; Sudak, L.J. Synergistic effect of hybrid stainless steel fiber and carbon nanotube on mechanical properties and electromagnetic interference shielding of polypropylene nanocomposites. *Compos. Part B* **2019**, *165*, 662–670. [CrossRef]
19. Kwon, S.; Ma, R.; Kim, U.; Choi, H.R.; Baik, S. Flexible electromagnetic interference shields made of silver flakes, carbon nanotubes and nitrile butadiene rubber. *Carbon* **2014**, *68*, 118–124. [CrossRef]
20. Lin, M.; Gai, Y.; Xiao, D.; Tan, H.; Zhao, Y. Preparation of pristine graphene paste for screen printing patterns with high conductivity. *Chem. Phys. Lett.* **2018**, *713*, 98–104. [CrossRef]
21. Jiang, Y.; Song, H.; Xu, R. Research on the dispersion of carbon nanotubes by ultrasonic oscillation, surfactant and centrifugation respectively and fiscal policies for its industrial development. *Ultrason. Sonochem.* **2018**, *48*, 30–38. [CrossRef] [PubMed]
22. Kirkpatrick, S. Percolation and Conduction. *Rev. Mod. Phys.* **1973**, *45*, 574–588. [CrossRef]
23. Cheng, H.-C.; Chen, C.-R.; Hsu, S.-H.; Cheng, K.-B. The electromagnetic shielding effectiveness of laminated fabrics using electronic printing. *Polym. Compos.* **2020**, *41*, 2568–2577. [CrossRef]
24. Choi, J.H.; Lee, K.Y.; Kim, S.W. Ultra-bendable and durable Graphene—Urethane composite/silver nanowire film for flexible transparent electrodes and electromagnetic-interference shielding. *Compos. Part B Eng.* **2019**, *177*, 107406. [CrossRef]
25. Leterrier, Y. 1-Mechanics of curvature and strain in flexible organic electronic devices. In *Handbook of Flexible Organic Electronics: Materials Manufacturing and Applications*; Woodhead Publishing: Oxford, UK, 2015; Volume 3, pp. 3–36.
26. Hewlett-Packard Co. *HP 8752 Operation Manual, Performance Tests*; Hewlett-Packard Co.: Palo Alto, CA, USA, 1996; Volume 4.
27. Cheng, K.-B.; Ramakrishna, S.; Ueng, T.H.; Lee, M.L. Electromagnetic Shielding Effectiveness of the Stainless Steel/Polyester Woven Fabrics. *Text. Res. J.* **2001**, *71*, 42–49. [CrossRef]
28. Hes, L.; Araujo, M.D.; Djulay, V.V. Effect of Mutual Bonding of Textile Layers on Thermal Insulation and Thermal Contact Properties of Fabric Assemblies. *Text. Res. J.* **1996**, *66*, 245–250. [CrossRef]
29. Banerjee, D.; Chattopadhyay, S.K.; Tuli, S. Infrared thermography in material research—A review of textile applications. *Indian J. Fibre Text. Res.* **2013**, *38*, 427–437.

sustainability

MDPI

Article

Geopolymer Technologies for Stabilization of Basic Oxygen Furnace Slags and Sustainable Application as Construction Materials

Wei-Hao Lee [1,*], Ta-Wui Cheng [1], Kuan-Yu Lin [1], Kae-Long Lin [2], Chia-Cheng Wu [3] and Chih-Ta Tsai [4]

1 Institute of Mineral Resources Engineering, National Taipei University of Technology, Taipei 10608, Taiwan; twcheng@ntut.edu.tw (T.-W.C.); dick89716245@gmail.com (K.-Y.L.)
2 Department of Environmental Engineering, National I-Lan University, I-Lan 260007, Taiwan; kllin@niu.edu.tw
3 CHC Resources Corporation, Kaohsiung City 80661, Taiwan; havanacj@chc.com.tw
4 Sustainable Environment Research Laboratories, National Cheng Kung University, Tainan City 701, Taiwan; chihta.tsai@gmail.com
* Correspondence: glowing955146@hotmail.com

Received: 15 May 2020; Accepted: 9 June 2020; Published: 18 June 2020

Abstract: The basic oxygen furnace slag is a major waste by-product generated from steel-producing plants. It possesses excellent characteristics and can be used as a natural aggregate. Chemically, the basic oxygen furnace slag encloses free CaO and free MgO, which is the main reason for the expansion crisis since these free oxides of alkaline earth metals react with water to form their hydroxide yields. The objective of the present research study is to stabilize the basic oxygen furnace slag by using innovative geopolymer technology, as their matrix contains a vast quantity of free silicon, which can react with free CaO and free MgO to form stable silicate compounds resulting in the prevention of the basic oxygen furnace slag expansion predicament. Lab-scale and ready-mixed plant pilot-scale experimental findings revealed that the compressive strength of fine basic oxygen furnace slag-based geopolymer mortar can achieve a compressive strength of 30–40 MPa after 28 days, and increased compressive strength, as well as the expansion, can be controlled less than 0.5% after ASTM C151 autoclave testing. Several pilot-scale cubic meters basic oxygen furnace slag-based geopolymer concrete blocks were developed in a ready-mixed plant. The compressive strength and autoclave expansion test results demonstrated that geopolymer technology does not merely stabilize the basic oxygen furnace slag production issue totally, but also turns the slags into value-added products.

Keywords: basic oxygen furnace slag; autoclave test; geopolymer technology; expansion behavior; recycling

1. Introduction

The Basic Oxygen Furnace (BOF) slag is a left-behind residual waste generated from the steelmaking industry. There are 1.5 million tons of BOF slags produced annually in Taiwan [1]. This high quantity leads to environmental and ecological issues together with health concerns. This waste demonstrates excellent physical and mechanical attributes, such as far-above-the-ground hardness, a higher compressive strength, and a near-to-the-ground abrasion ratio. It is capable of substituting natural aggregates to produce structural materials, road pavements, and more. Unfortunately, the key setback of BOF slags is their expansion behavior because of their gigantic content of free lime, as a consequence, a hindrance in the context of BOF slag recycling and reuse takes place [2–6]. Most of the coarse BOF slags, larger than four mesh, are mixed with asphalt to produce asphalt-concrete which can be employed as road pavements. However, approximately 0.7 million tons of fine BOF slags, which is less than four mesh, are unable to be utilized effectively, especially used in the Portland cement system [7]. For that reason, the way in which a BOF slag can be reutilized as a fine aggregate has become an imperative issue.

Previous studies on BOF slags used as aggregates in concrete and their advantageous properties have been extensively reported [4,5,8–11]. However, the BOF slag contains free-CaO and free-MgO that can result in volumetric instability due to their reaction with water, yielding alkaline earth metal hydroxides. This must be addressed, employing appropriate treatment to prevent expansion. There are several ways to stabilize BOF slags, such as stabilization in the hot or molten stage [12–14], and stabilization by employing a carbonation process [12,15,16]. Singh et al. supplemented oxalic acid into cement mortar to improve the densification and mechanical properties of concrete [17], while Ding et al. utilized scrubbing attrition and chelating reagent treatment to remove free lime from BOF slags [5].

On the other hand, Lin et al. immersed BOF slags into the water to provide sufficient stabilization reactions [18]. According to previous research results, the adding up of silica sand in the molten stage can complete the solution and get rid of free lime crisis. This brings into question whether or not analogous reactions can occur at room temperature. If possible, they will entirely resolve the quandary of BOF slag expansion and achieve the goals of waste recycling following the principles of the Circular Economy.

Nowadays, innovative geopolymer technology is drawing the attention of researchers on account of not only its nine-times lower carbon footprint and six-times lower energy consumption as compared to the OPC (Ordinary Portland Cement) system [19], but also due to the demonstration of excellent properties by geopolymer composites, such as high initial strength, brilliant resistance to chemicals and freeze-thaw as well as thermal and fire calamities, sustainability, and strength and durability. This means simply that novel geopolymer technology can lend a hand in solving the great dilemma of global warming and the saving of natural resources of rocks and minerals. On top of that, it is cost-effective, since it can incorporate profusely accessible various wastes in the manufacture of geopolymer composites, which otherwise create problems of landfilling and pollution of the environment, soil, surface and subsurface waters besides health hazards [20,21]. Accordingly, it also extends to an organized solution of the disposal of these wastes. Inorganic geopolymers are analogous to natural zeolite minerals but possess an amorphous microstructure, they are classed as three-dimensionally networked materials synthesized through the reaction of rich aluminosilicate materials, such as precursors and alkaline solutions, as activators [22–24]. During the geopolymerization reaction, Si gel and Al gel were produced on the solid particle surface, and thus, they formed the Si-O-Al framework. SiO_4 and AlO_4 tetrahedra are linked to each other by sharing all O_2 atoms [25,26]. Their unique internal structure is the core reason for why geopolymer has superior chemical and physical properties such as high compressive strength, high durability, favorable structural integrity, and low permeability. These exothermic and complex geopolymer processes contain large amounts of free silicon. This free silicon can react with oxides of free lime and free magnesium in the BOF slag, and hence, the formation of stable compounds takes place, thereby inhibiting the expansion of the BOF slag.

In this original research, BOF slags, granulated blast furnace slag (GGBS) powder, coal fly ash (FA) and alkali activator solutions were employed as raw materials to manufacture BOF slag-based geopolymer mortar. The effect on compressive strength and expansion behavior from the GGBS/FA ratio and the SiO_2/Na_2O ratio in geopolymer mortar are presented in this investigation. For testing the expansion behavior of BOF slag-based geopolymer mortar, the application of the autoclave testing method as the accelerated test was made [27]. Lab-scale and ready-mixed plant pilot-scale recycling processes are also investigated in this study.

2. Assumption of the Mechanism for Stabilization BOF Slags in Geopolymer System

In general, the expansion phenomenon of the BOF slag is mainly due to the f-CaO, and f-MgO content presented. When the BOF slag makes contact with water, f-CaO will undergo a hydration reaction, causing volume expansion. The f-MgO hydration reaction is relatively slow, and about 30 days later, it will also cause volume expansion. After the BOF slag absorbs water, f-CaO will become calcium hydroxide "$Ca(OH)_2$", the volume will expand by 127% [28], and f-MgO will become magnesium hydroxide "$Mg(OH)_2$". The volume expansion is about 118% [29]. Therefore, this study conceived that utilizing free silicon to react f-CaO into a stable calcium silicate compound would prevent the expansion. The relevant reaction Equations (1) and (2) are as follows:

$$\text{f-CaO} + \text{Si} \rightarrow \text{CaSiO}_3 \quad (1)$$

$$\text{f-MgO} + \text{Si} \rightarrow \text{MgSiO}_3 \quad (2)$$

After the geopolymer reaction is completed, the excessively free silicon will remain in the matrix of BOF slag-based geopolymer. If other factors caused the geopolymer matrix and/or BOF slag aggregates to be broken, the f-CaO may be released again. At this time, if water infiltrates from the crack, the free silicon remaining in the geopolymer product will dissolve in the water, and then react with f-CaO to form a stable calcium silicate to prevent expansion. The reaction schematic diagram is shown in Figure 1.

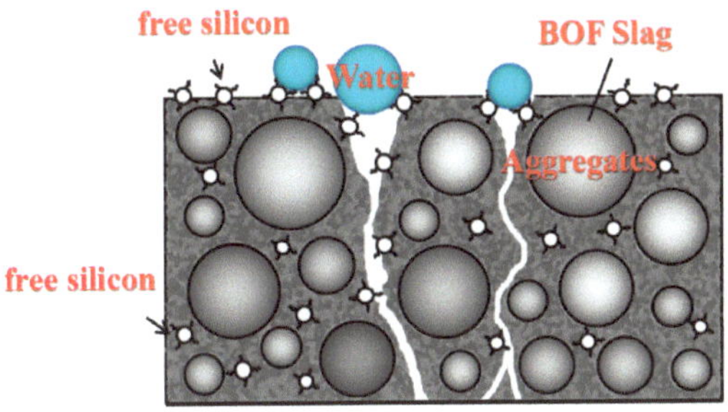

Figure 1. Schematic diagram of f-CaO reaction with free silicon in geopolymer matrix.

3. Experimental

3.1. Materials

BOF slag aggregates with a particle size less than four mesh original fine with a D_{50} at 0.75 mm obtained from CHC Resources Corporation were utilized. The chemical composition of fine BOF slags is depicted in Table 1. The key composition in BOF slag was CaO and Fe_2O_3; it also contains 4.7 wt% of free lime. Additionally, GGBS and FA were obtained from CHC Resources Corporation in Taiwan. GGBS possesses a particle size between 0.6–135.7 μm with D_{50} at 12.3 μm while those of FA range among 0.7–201.9 μm with a D_{50} at 22.2 μm. The chemical composition of GGBS powder and FA were also represented in Table 1. Alkali activator solutions with various SiO_2/Na_2O molar ratios were prepared by mixing sodium silicate solution (9.5 wt% Na_2O, 29 wt% SiO_2) and 6M sodium hydroxide. The molar ratio for SiO_2/Al_2O_3 was kept at 50 and was controlled by sodium aluminate. The alkali solutions were prepared and ready to use the day before the experiment.

Table 1. Chemical composition of basic oxygen furnace (BOF) slag, granulated blast furnace slag (GGBS) and coal fly ash (FA).

Composition / wt.%	SiO_2	CaO	Al_2O_3	Fe_2O_3	MgO	f-CaO	LOI.	Others
BOF Slag	9.4	37.1	4.2	24.1	7.1	4.7	0.8	12.6
GGBS	27.7	57.9	11.2	0.4	–	–	–	2.8
FA	60.2	2.7	19.1	8.7	–	–	2.9	6.4

3.2. Methods

GGBS and FA were mixed at the designed weight ratio with an additional 3 wt% of wollastonite added during blending. Subsequent to pre-mixing for 3 min, the mixture was then activated with alkali solutions. After thorough mixing for 3 min, the fine BOF slags were then added into geopolymer paste for an additional 3 min blending to prepare geopolymer mortar with various geopolymer: BOF slag ratios. The geopolymer mortar was then cast into cylindrical molds of size (Φ50 × 100 mm). Following 24 h of de-molding, the samples were then cured at room temperature until certain days for testing. In order to measure the volume stability, all the BOF slag-based geopolymer mortar samples were tested using the autoclave testing method according to ASTM C151 standard.

Before and after autoclave testing, the length, diameter, and volume change rate of the cylindrical specimen were calculated by dividing the circle portion of the cylindrical specimen into three sections with the center point of the circle. A Vernier caliper was used for measuring the length varieties of the cylindrical sample. For diameter measurements, an arbitrarily selected marked two points at the middle of the height of the cylindrical test body was measured and another two points were made at a position rotated by 90°, which were also measured. The change in total volume is calculated from the above results.

4. Results and Discussion

4.1. BOF Slags Expansion Behavior and Powdering Rate

The powdering rate measurement was based on GB/T 24175-2009 (Test method for stability of steel slag). Before collecting an 800 g oversize part to put into an autoclave, the BOF slags were screened using a No. 4 sieve. The autoclave maintained the pressure at 20.8 ± 0.7 kgf/cm^2, temperature 215.7 ± 1.7 °C for 3 h. Subsequently after drying and sieving again by employing the No. 4 sieve, the powdering rates were calculated as illustrated in Equation (3).

Table 2 demonstrates the results of powdering rates for various size ranges of BOF slags after autoclave tests. The powdering rate increased from 17.4% to 29.9% with decreasing particle size ranges from >3/4 inch–3/8 inch–No. 4 mesh. This implies that the finer particle size range of BOF slags contained more free lime.

Table 2. Powdering rate for different size range BOF slags after autoclave test.

Particle Size Range	Powdering Rate (%)
3/8 inch–#4 mesh	29.9
3/4 inch–3/8 inch	27.8
>3/4 inch	17.4

$$\text{Powdering Rate} = \frac{Weight\ of\ less\ than\ 4\ mesh\ part}{Total\ Sample\ weight} \times 100\% \quad (3)$$

In order to compare the performance of BOF slags (aggregate) in the Portland cement system and geopolymer system (binder), the binder/aggregate weight ratio was kept at 1:2.75 for the BOF slag expansion test. The expansion behavior of the BOF slag in the Portland cement system and geopolymer technology is presented in Table 3. Based on the experiment results, it is seen that in the case of the BOF slag in the Portland cement system, the sample totally collapsed after the autoclave test due to the steam, which accelerated the reaction of free lime, thus causing expansion. Nevertheless, in the case of the BOF slag in the geopolymer system, the sample still maintained its integrity after the autoclave test as displayed in Table 3. The average height, diameter, and volume changes recorded are 0.18%, 0.07%, and 0.35%, respectively. After the autoclave test, the samples were crushed, grinded and analyzed by XRD (X-ray Diffraction) (Figure 2). The mineral phase of calcium silicate was found, indicating that the geopolymer technique can stabilize the untreated BOF slag fines. This is ascribed to the enormous quantities of free silicon present in the geopolymer matrix. This free silicon reacted with free lime or free-MgO on the BOF slag surface to form a stable compound. When the BOF slag-based geopolymer is subjected to an external force and cracks crop up, the moisture will enter into the BOF slags. The free silicon in the geopolymer matrix will be dissolved and brought into internal BOF slag to react with free-CaO or free-MgO to form stable calcium silicate or magnesium silicate. This reaction can effectively inhibit the expansion of the BOF slags.

Table 3. BOF slag expansion behavior in Portland cement system and geopolymer system after autoclave test.

BOF Slag in Different System	Before Autoclave Test	After Autoclave Test
Portland Cement System	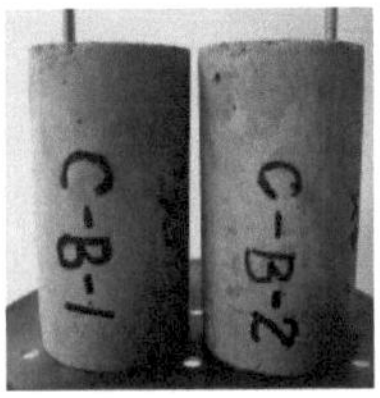	
Geopolymer System		
Ave height change	0.18%	
Ave diameter change	0.07%	
Ave volume change	0.35%	

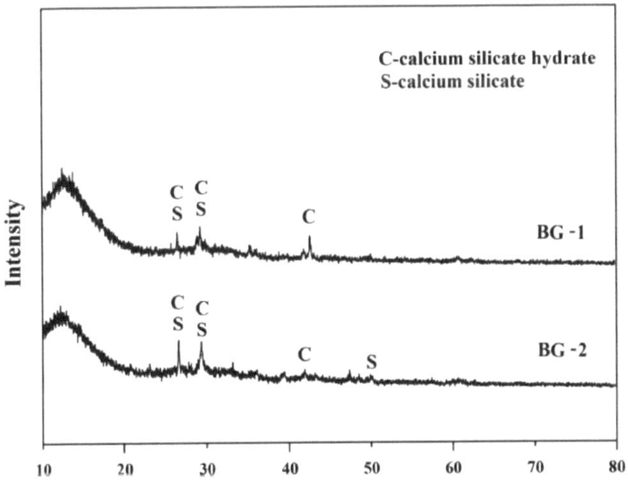

Figure 2. Results of XRD (X-ray Diffraction) analysis of the BOF slags in the geopolymer system after autoclave test.

4.2. Effect of SiO_2/Na_2O Molar Ratio on the Properties of BOF Slag-Based Geopolymer Mortar

In order to understand the effect of the SiO_2/Na_2O molar ratio on the features depending on the BOF slag-based geopolymer mortar, the experimental parameters were adjusted, and the experimental results exhibit that the compressive strength depending on the SiO_2/Na_2O ratio of the BOF slag-based geopolymer mortar is augmented as the ratio of SiO_2/Na_2O increases, as portrayed in Figure 3. Both tend to enhance the compressive strength as the curing time increases. When the ratio of SiO_2/Na_2O was between 1.28 and 1.5, the compressive strength at the age of 56 days was about 32–38 MPa. Likewise, when the ratio of SiO_2/Na_2O was 1.6, the compressive strength achieved 53 MPa on the day 56. However, only the decrease of compressive strength after 56 days for SiO_2/Na_2O = 1.4 was found. The reason for this is large amounts of micropores generated in the structure of the geopolymer that hinder the development of compressive strength, and no such phenomenon is found in a ratio higher or lower than the SiO_2/Na_2O ratio of 1.4 [30]. Moreover, it is found that most of the specimens subjected to the autoclave expansion treatment according to the ASTM C151 standard possess a tendency to increase the compressive strength, and the strength on day 56 can obtain 40–55 MPa as illustrated in Figure 3b. A careful observation of the surface of the specimen after the autoclave test revealed that it still has a small amount of surface peeling, which may be due to the fact that the BOF slag particles on the surface of the specimen cannot be entirely covered by the geopolymeric slurry, or maybe because of the highly thin coating. Therefore, there is still a small part of the reaction expansion phenomenon. However, due to its substantial increase in compressive strength, it indicates that the overall performance of the test body subsequent to the autoclave expansion test is still very stable.

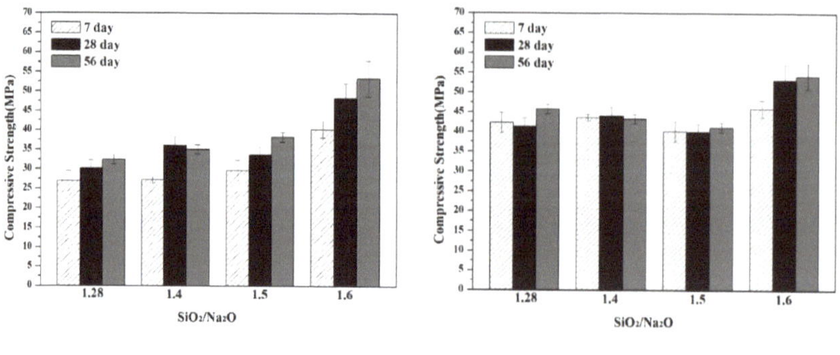

(a) (b)

Figure 3. Effect of SiO_2/Na_2O molar ratio on the curing age and compressive strength of BOF slag-based geopolymer mortar before autoclave expansion test (**a**) and after autoclave expansion test (**b**).

The effect of the SiO_2/Na_2O molar ratio on the curing age of the BOF slag-based geopolymer mortar was subjected to an autoclave expansion test specimen, and its linear expansion, diameter expansion, and bulk expansion characteristics were analyzed. The results are shown in Figure 4. As illustrated in Figure 4, the volume expansion ratio of each of the samples was 0.4% or less, of which, the expansion ratio of SiO_2/Na_2O = 1.4 and 1.5 is the lowest and is only 0.1% or less. Furthermore, the linear expansion ratio and the diameter expansion ratio were analyzed ahead, and the linear expansion ratio was mostly less than 0.1%, and the diameter expansion ratio was also 0.15% or less. This shows that the untreated BOF slag has high stability under the geopolymer system. According to the results of this experiment, the subsequent selection of the BOF slag-based geopolymer mortar was carried out with a ratio of SiO_2/Na_2O of 1.5.

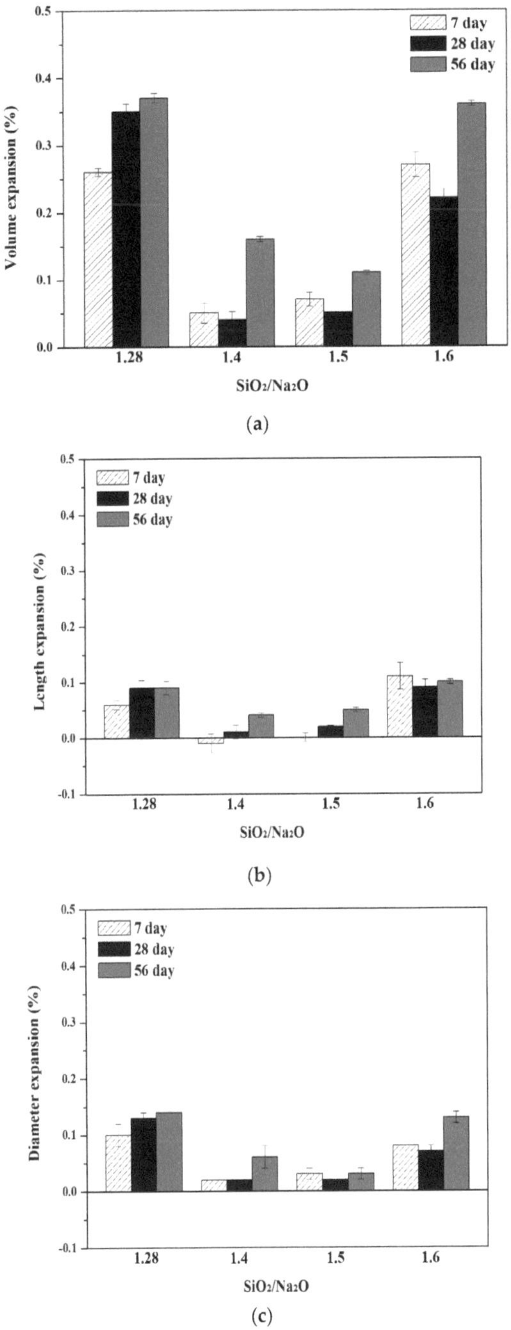

Figure 4. Effect of SiO_2/Na_2O molar ratio on the length, diameter, and volume changes after autoclave test, (**a**) volume expansion ratio (**b**) linear expansion ratio (**c**) diameter expansion ratio.

4.3. *Effect of GGBS/FA Ratio on the Properties of BOF Slag-Based Geopolymer Mortar*

The results of the compressive strength of the BOF slag-based geopolymer mortar, with varying proportions of the GGBS and FA, are shown in Figure 5. The compressive strength increased with increased GGBS content (Figure 5a). This could be due to the fact that the structure of geopolymer mortar is denser and more increased with curing age. The BOF slag-based geopolymer mortar of different powder ratios was subjected to a compressive strength test after an autoclave treatment, and the results are shown in Figure 5b. According to the experimental results, it can be found that although the surface of the test piece has a small part of peeling after the autoclave test, the compressive strength is still significantly increased. This means that the test specimen is still very stable after the autoclave expansion test. The higher the content of fly ash, the significantly higher the strength of the sample after the autoclave test. This may be due to the high silicon content of fly ash in the system to suppress the BOF slag expansion [31].

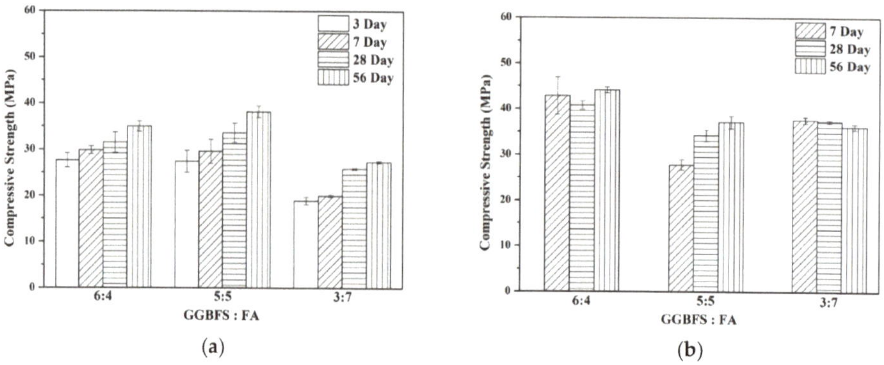

Figure 5. Effect of GGBS/FA ratio on compressive strength of BOF slag-based geopolymer mortar (**a**) before autoclave expansion test; (**b**) after autoclave expansion test.

The autoclave expansion test, ASTM C151, was carried out to understand the stability and volume expansion rate of BOF slag-based geopolymer mortar samples, with different proportions of GGBS:Fly ash, shown in Table 4. Table 4 highlights that in the BOF slag-based geopolymer mortar sample, the excessive GGBS or the excessively high FA content has a high expansion rate. Moreover, with a ratio of GGBS:FA = 6:4, the volume expansion rate is 0.53% after 56 days of curing, but the GGBS:FA = 3:7, after 56 days of curing, has an expansion rate as high as 0.71%. Only between the GGBS:FA = 5:5, an expansion rate of less than 0.1% exists. The linear expansion ratio and the diameter expansion ratio were less than 0.3%.

Table 4. The effect of GGBS/FA Ratio on the expansion ratio of BOF slag-based geopolymer mortar after autoclave test.

GGBS: FA	Diameter Expansion (%)			Length Expansion (%)			Volume Expansion (%)		
	7d	28d	56d	7d	28d	56d	7d	28d	56d
6:4	0.18	0.18	0.18	0.24	0.22	0.15	0.62	0.59	0.53
5:5	0.03	0.02	0.02	0.01	0.02	0.03	0.07	0.05	0.10
3:7	0.24	0.21	0.26	0.16	0.19	0.19	0.66	0.63	0.71

Similarly, in the ratio of GGBS:FA = 5:5, both the diameter and linear expansion rate were less than 0.03%. Although the FA in the system can provide more silicon to inhibit the expansion of the BOF slag, the powder itself has large shrinkage and insufficient strength, thus cannot add too much. On the other hand, in the sample with higher GGBS content, although the content of silicon—which can react with free lime—is reduced, it is still sufficient to inhibit the expansion of the BOF slag due to its own strength, and its volume expansion is still 0.53%. In the GGBS:FA = 5:5, the strength provided by the GGBS in the system and the FA that can react to inhibit free calcium reached the optimum amount, and consequently the expansion rate and age are the best, <0.1% for 7–56 days curing.

4.4. Laboratory Horizontal Double Shaft Mixer Tests

This phase of the test is mainly to simulate the large-scale test of the actual plant. GGBS and FA (5:5) are used as source materials. The alkaline liquid has NaOH concentration of 6M, the SiO_2/Na_2O molar ratio is 1.5, the SiO_2/Al_2O_3, as well as molar ratio, is 50. The moisture content of the BOF slag is controlled at 10%, where the BOF slag is not pretreated before the tests. To simulate the actual factory test, this experiment prepared a BOF slag-based geopolymer mortar with a horizontal biaxial mixing machine and poured a test specimen of Φ10 cm×20 cm for the compression test and the autoclave expansion test. The mixture proportion and process are shown in Table 5. Table 6 shows the hardening time of the BOF slag-based geopolymer mortar. According to the hardening time test results, the initial setting time is about 3 h; the final setting time is about 9 h.

Table 5. Simulation the large-scale experiment for BOF slag-based geopolymer mortar.

Mix No.	Source Materials		Binder: Aggregate	Wollastonite	L/S	BOF Slags Water Content
	FA	GGBS				
GC-BOF10%	5	5	1:2.75	5%	0.49	10%

Table 6. Hardening time result of simulation of large-scale experimental for BOF slag-based geopolymer mortar.

Mix No.	Setting Time	
	Initial	Final
GC-BOF10%	3 h 15 min	8 h 40 min

The compressive strength of the simulated large-scale experimental BOF slag-based geopolymer mortar is shown in Figure 6. As the curing age increases, the strength of the test body increases. At seven days of age, the strength reached about 27 MPa; at 28 days of age, the strength reached 40 MPa. It is said that the simulation of the large-scale experimental of BOF slag-based geopolymer mortar has excellent strength performance.

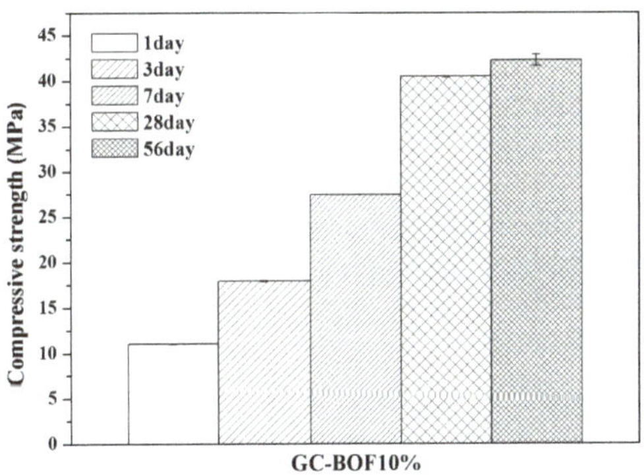

Figure 6. The compressive strength of simulation large-scale experimental BOF slag-based geopolymer mortar.

The autoclave expansion test of the simulated large-scale experimental BOF slag-based geopolymer mortar is shown in Table 7. After the autoclave expansion test, the BOF slag-based geopolymer mortar sample is complete, and only a slight surface is peeling. The compressive strength test before and after autoclave found that the surface peeling did not affect the stability of the test specimen.

Table 7. Autoclave expansion test (seven days) results of simulating the large-scale of the experimental BOF slag-based geopolymer mortar.

	Before Autoclave Test	After Autoclave Test	Volume Changed
Test Specimen GC-BOF10%			0.27%
Compressive Strength (MPa)	28.4	35.6	

4.5. BOF Slag-Based Geopolymer Mortar Tests in Ready-Mixed Plant

4.5.1. Ready-Mixed Plant Small Scale Tests

The mixed proportion for Ready-mixed plant small scale test is shown in Table 8. The main difference between the two tests is the ratio of binder and aggregate, which is the additional amount of the BOF slags.

Table 8. Mixture proportion of ready-mixed plant small-scale test.

NO.	L/S	Binder:Aggregate	Mixture Proportion (kg)				Total Weight (kg)
			GGBS	FA	BOF (15% water)	Alkali Solution	
Test-1	0.41	1:2.90	15.12	15.12	100.85 (87.70)	12.50	143.59
Test-2	0.38	1:1.98	19.70	19.70	90.00 (78.26)	15.00	144.40

The fresh properties of the ready-mixed plant small-scale test mixture are shown in Table 9. According to the results, it is found that the BOF slags of the high-water content of BOF slag-based geopolymer mortar showed excellent workability. The initial slump flow of Test-1 is 540*580 mm, and it is allowed to stand for 45 min after, and the slump flow is reduced to only 520*520 mm. Test-2's initial slump flow is 470*480 mm, and after allowing it to stand for 45 min, its slump flow is reduced to 440*440 mm. The results of the two tests were found to have low slump flow loss performance.

Table 9. Fresh properties mixture after ready-mixed plant small scale test.

NO.	Initial		After 45 Min	
Test-1	Slump 265 mm	Slump flow 540*580 mm	Slump 250 mm	Slump flow 520*520 mm

Test-2	Initial		After 45 min	
	Slump 260 mm	Slump flow 470*480 mm	Slump 250 mm	Slump flow 440*440 mm
	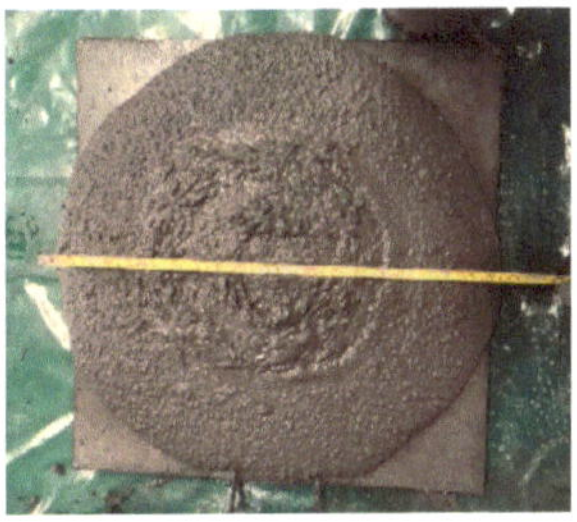			

The compressive strength of the ready-mixed plant small scale tests is shown in Table 10 and Figure 7. As the curing age increases, the strength of the test body increases. The strength of the late Test-2 is higher than that of Test-1 because the strength source of the geopolymer system is from the geopolymer slurry rather than an aggregate. Therefore, in Test-2, where the BOF fine aggregate is relatively low, the compressive strength will be higher than Test-1. However, after the autoclave test, the compressive strength decreases in Test-2. The reason for this is due to water release and formed cracks.

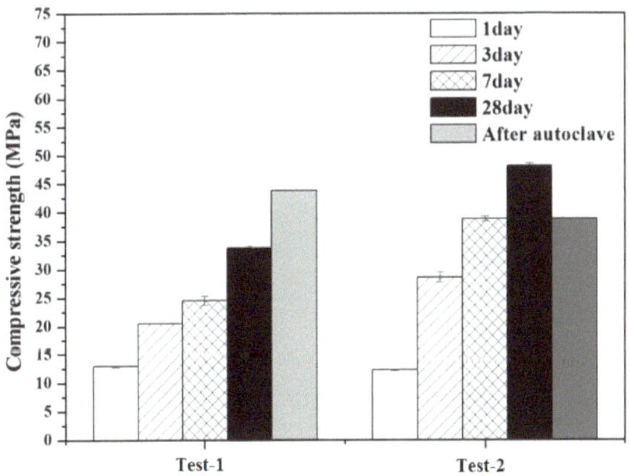

Figure 7. Compressive strength of the ready-mixed plant small scale test.

Table 10. Compressive strength of concerning plant small scale test.

NO.	Test Time	Compressive Strength (MPa)								
		1d	Ave	3d	Ave	7d	Ave	28d	Ave	After Autoclave Test
Test-1	0 min	13.2	12.9	20.4	20.5	22.9	24.5	32.9	33.7	43.8
	45 min	12.7		20.5		26.1		34.5		
Test-2	0 min	12.1	12.2	26.8	28.6	39.7	38.8	49.3	48.1	38.8
	45 min	12.3		30.4		38.0		47.0		

4.5.2. Ready-Mixed Plant Pilot-Scale Tests

Ready-mixed plant pilot-scale test parameters are shown in Table 11. The ratios of FA and GGBS are 5:5 and 6:4, the ratios of binder and aggregate are 1:2.936 and 1:3.575. The total test volume is 1.5 cubic meters, and the total weight in each test is approximately 3.5–3.6 tons.

Table 11. Mixture proportion of ready-mixed plant pilot-scale test.

NO.	FA : GGBS	L/S	Binder : Aggregate	Mixture Proportion (kg)						Total Volume and Weight
				Alkali Solution	GGBS	FA	BOF Slag (Wet)	BOF Slag Water Content	Water Added	
Test-3	5:5	0.50	1:2.936	374	375	375	2,400	9.0%	20.0	1.5 m^3 3544 kg
Test-4	6:4	0.52	1:3.575	349	267	400	2,560	7.4%	60.0	1.5 m^3 3636 kg

The fresh properties of the ready-mixed plant pilot-scale tests are shown in Table 12. The test number of Test-3 and Test-4 have a slump flow of 380*390 mm and 510*490 mm, respectively. The reason for this difference is the increase in the use of FA in Test-4 samples, which is spherical in shape and contributes to its fluidity [32,33].

Table 12. Fresh properties of ready mixed plant pilot-scale tests.

NO.	FA:GGBS	Slump	Slump Flow
Test-3	5:5	 260 mm	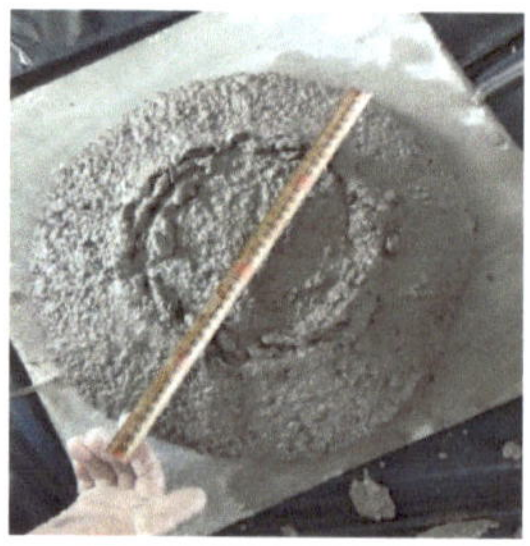 380*390 mm
Test-4	6:4	 270 mm	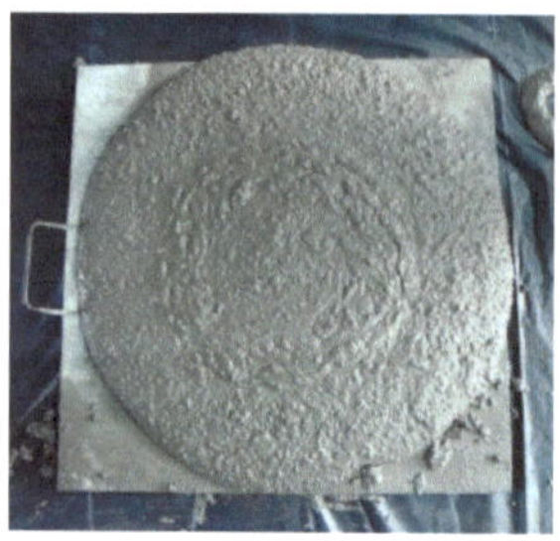 510*490 mm

The compressive strength and their autoclave test of ready-mixed plant pilot-scale experiments are shown in Table 13 and Figure 8. As the curing age increases, the strength of the test body increases. The compressive strength of Test-3 is slightly higher than that of Test-4. The main reason is the adjustment of the ratio of FA to GGBS. Test-4 is higher in the amount of FA, and the reactivity of FA itself is poorer than that of GGBS powder which causes its intensity to be slightly lower than Test-3. According to the results of the autoclave expansion test, as the curing age increases, the structure of the test body is more complete, and the test specimen has no break point after the autoclave expansion test. The expansion changes after the autoclave test for the ready-mixed plant pilot-scale tests are shown in Table 14. All the expansion test results can be controlled around −0.41% for diameter and liner changes.

Table 13. Compressive strength and autoclave test of ready-mixed plant pilot-scale tests.

NO.	FA: GGBS	Compressive Strength (MPa)								Autoclave Test
		1d	Ave	3d	Ave	7d	Ave	28d	Ave	28d
Test-3	5:5	20.3 21.5	20.9	33.9 32.8	33.4	37.2 37.9	37.5	39.7 43.4	41.6	45.4
Test-4	6:4	13.8 14.1	13.9	24.9 26.0	25.4	31.6 31.1	31.4	39.7 40.4	39.9	31.7

Table 14. Expansion changes after autoclave test for ready-mixed plant pilot-scale tests (curing time 28 days).

NO.	Expansion Ratio After Autoclave Test (%)	
	Diameter Change	Length Change
Test-3	−0.41	−0.41
Test-4	−0.41	−0.41

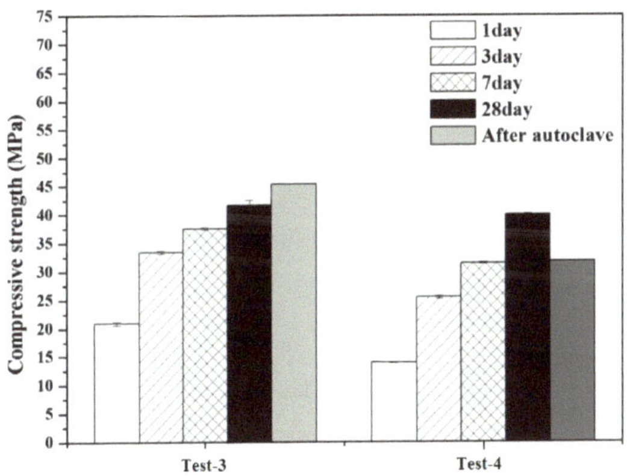

Figure 8. Compressive strength of ready-mixed plant pilot-scale test.

5. Conclusions

The present study and its findings were piloted, and conclude that:

- Reduction concerning CO_2 emissions and consumption of copiously available BOF slag wastes is possible by employing geopolymer technology. Not merely that, it stabilizes the BOF slag production absolutely by turning them into valuable products.
- Stabilization of the BOF slag through geopolymer technology is successfully determined, since their matrix encompasses a large amount of free silicon (Si) which can react with free CaO and free MgO to form stable silicate compounds leading to the addressing of the quandary of BOF slag expansion.

- Lab-scale and ready-mixed plant pilot-scale experimental upshots unveiled that the compressive strength of fine BOF slag-based geopolymer mortar achieved the compressive strength of 30–40 MPa after 28 days, and increased compressive strength. The expansion can be controlled also, less than 0.5% after ASTM C151 autoclave testing.
- A systematic solution for the disposal of the waste accumulation of BOF slags is extended through this novel geopolymer technology since its incorporation is possible in manufacturing, user and eco-friendly green geopolymer composites, otherwise filling land spaces and causing the contamination of environments, ecology, soils, surface and subsurface waters as well as health hazards.

Author Contributions: Conceptualization, W.-H.L. and T.-W.C.; methodology, T.-W.C.; C.-C.W.; C.-T.T. and W.-H.L.; investigation, W.-H.L.; K.-Y.L.; C.-C.W.; C.-T.T. and T.-W.C.; writing—original draft preparation, W.-H.L. and T.-W.C.; writing—review and editing, W.-H.L. and T.-W.C.; supervision, K.-L.L. and T.-W.C.; project administration, T.-W.C and K.-L.L.; funding acquisition, T.-W.C. All authors have read and agreed to the published version of the manuscript.

Funding: This research was funded by CHC Resources Corporation, the Ministry of Science and Technology, Taiwan, contract number MOST 106-3114-E-027-001 & MOST 107-2218-E-027-006.

Conflicts of Interest: The authors declare no conflict of interest.

References

1. Wang, C.C. Engineering Characteristics of Basic Oxygen Furnace Slag and Used as Base Layer Material. Ph.D. Thesis, National Cheng Kung University, Taiwan, 2005.
2. Shi, C. Steel slag—Its production, processing, characteristics, and cementitious properties. *J. Mater. Civ. Eng.* **2004**, *16*, 230–236. [CrossRef]
3. Wang, G. Determination of the expansion force of coarse steel slag aggregate. *Constr. Build. Mater.* **2010**, *24*, 1961–1966. [CrossRef]
4. Brand, A.S.; Roesler, J.R. Steel furnace slag aggregate expansion and hardened concrete properties. *Cem. Concr. Compos.* **2015**, *60*, 1–9. [CrossRef]
5. Ding, Y.C.; Cheng, T.W.; Liu, P.C.; Lee, W.H. Study on the treatment of BOF slag to replace fine aggregate in concrete. *Constr. Build. Mater.* **2017**, *146*, 644–651. [CrossRef]
6. Kambole, C.; Paige-Green, P.; Kupolati, W.K.; Ndambuki, J.M.; Adeboje, A.O. Basic oxygen furnace slag for road pavements: A review of material characteristics and performance for effective utilisation in southern Africa. *Constr. Build. Mater.* **2017**, *148*, 618–631. [CrossRef]
7. Lu, T.H.; Chen, Y.L.; Shih, P.H.; Chang, J.E. Use of basic oxygen furnace slag fines in the production of cementitious mortars and the effects on mortar expansion. *Constr. Build. Mater.* **2018**, *167*, 768–774. [CrossRef]
8. Blanco, I.; Molle, P.; Saenz de Miera, L.E.; Ansola, G. Basic Oxygen Furnace steel slag aggregates for phosphorus treatment. Evaluation of its potential use as a substrate in constructed wetlands. *Water Res.* **2016**, *89*, 355–365. [CrossRef]
9. Bodor, M.; Santos, R.M.; Cristea, G.; Salman, M.; Cizer, O.; Iacobescu, R.I.; Chiang, Y.W.; van Balen, K.; Vlad, M.; van Gerven, T. Laboratory investigation of carbonated BOF slag used as partial replacement of natural aggregate in cement mortars. *Cem. Concr. Compos.* **2016**, *65*, 55–66. [CrossRef]
10. Chen, S.H.; Lin, D.F.; Luo, H.L.; Lin, Z.Y. Application of reclaimed basic oxygen furnace slag asphalt pavement inroad base aggregate. *Constr. Build. Mater.* **2017**, *157*, 647–653. [CrossRef]
11. Amelian, S.; Manian, M.; Abtahi, S.M.; Goli, A. Moisture sensitivity and mechanical performance assessment of warm mix asphalt containing by-product steel slag. *J. Clean. Prod.* **2018**, *176*, 329–337. [CrossRef]
12. Santos, R.M.; Ling, D.; Sarvaramini, A.; Guo, M.; Elsen, J.; Larachi, F.; Beaudoin, G.; Blanpain, B.; Van Gerven, T. Stabilization of basic oxygen furnace slag by hot-stage carbonation treatment. *Chem. Eng. J.* **2012**, *203*, 239–250. [CrossRef]
13. Yin, X.; Zhang, C.; Wang, G.; Yang, J.; Cai, Y.; Zhao, C. Free CaO stabilisation by mixing of BF slag and BOF slag in molten state. *Ironmak. Steelmak.* **2018**, *45*, 969–977. [CrossRef]
14. Yin, X.; Zhang, C.; Wang, G.; Cai, Y.; Zhao, C. Stabilization of free CaO in molten BOF slag by addition of silica at high temperature. *Metall. Res. Technol.* **2018**, *115*, 414. [CrossRef]

15. Pan, S.Y.; Chiang, P.C.; Chen, Y.H.; Tan, C.S.; Chang, E.E. Ex Situ CO_2 Capture by Carbonation of Steelmaking Slag Coupled with Metalworking Wastewater in a Rotating Packed Bed. *Environ. Sci. Technol.* **2013**, *47*, 3308–3315. [CrossRef] [PubMed]
16. Chen, K.W.; Pan, S.Y.; Chen, C.T.; Chen, Y.H.; Chiang, P.C. High-gravity carbonation of basic oxygen furnace slag for CO_2 fixationand utilization in blended cement. *J. Clean. Prod.* **2016**, *124*, 350–360. [CrossRef]
17. Singh, N.K.; Mishra, P.C.; Singh, V.K.; Narang, K.K. Effects of hydroxyethyl cellulose and oxalic acid on the properties of cement. *Cem. Concr. Res.* **2003**, *33*, 1319–1329. [CrossRef]
18. Lin, P.C.; Hsieh, H.C.; Ou, M.Y.; Huang, C.H.; Hsueh, W.H. Water Immersing and Aging Treatment of Steel-making Slag (CHC Resources Corp., Taiwan; China Steel Corp., Taiwan). Taiwan Patent TW-I275647, 11 March 2007.
19. Louati, S.; Hajjaji, W.; Baklouti, S.; Samet, B. Structure and properties of new eco-material obtained by phosphoric acid attack of natural Tunisian clay. *Appl. Clay Sci.* **2014**, *101*, 60–67. [CrossRef]
20. Ferone, C.; Capasso, I.; Bonati, A.; Roviello, G.; Montagnaro, F.; Santoro, L.; Turco, R.; Cioffi, R. Sustainable management of water potabilization sludge by means of geopolymers production. *J. Clean. Prod.* **2019**, *229*, 1–9. [CrossRef]
21. Capasso, I.; Lirer, S.; Flora, A.; Ferone, C.; Cioffi, R.; Caputo, D.; Liguori, B. Reuse of mining waste as aggregates in fly ash-based geopolymers. *J. Clean. Prod.* **2019**, *220*, 65–73. [CrossRef]
22. Swanepoel, J.C.; Strydom, C.A. Utilisation of fly ash in a geopolymeric material. *Appl. Geochem.* **2002**, *17*, 1143–1148. [CrossRef]
23. Boca Santa, R.A.A.; Soares, C.; Riella, H.G. Geopolymers with a high percentage of bottom ash for solidification/immobilization of different toxic metals. *J. Hazard. Mater.* **2016**, *318*, 145–153. [CrossRef] [PubMed]
24. Yaseri, S.; Hajiaghaei, G.; Mohammadi, F.; Mahdikhani, M.; Farokhzad, R. The role of synthesis parameters on the workability, setting and strength properties of binary binder based geopolymer paste. *Constr. Build. Mater.* **2017**, *157*, 534–545. [CrossRef]
25. Wang, W.C. Feasibility of stabilizing expanding property of furnace slag by autoclave method. *Constr. Build. Mater.* **2014**, *68*, 552–557. [CrossRef]
26. Zhang, Y.J.; Li, H.H.; Wang, Y.C.; Xu, D.L. Geopolymer Microstructure and Hydration Mechanism of Alkali-activated Fly Ash-based Geopolymer. *Adv. Mater. Res.* **2012**, *374–377*, 1481–1484. [CrossRef]
27. Davidovits, J. *Geopolymer Chemistry and Applications*; Geopolymer Institute: Saint-Quentin, France, 2008.
28. Wang, G.C. *The Utilization of Slag in Civil Infrastructure Construction*; Woodhead Publishing Series in Civil Engineering; Woodhead Publishing: Duxford, UK, 2016.
29. Ghosh, S.N. *Advances in Cement Technology: Critical Reviews and Case Studies on Manufacturing, Quality Control, Optimization and Use*; Elsevier: New Delhi, India, 2014.
30. Cho, Y.K.; Yoo, S.W.; Jung, S.H.; Lee, K.M.; Kwon, S.J. Effect of Na_2O content, SiO_2/Na_2O molar ratio, and curing conditions on the compressive strength of FA-based geopolymer. *Constr. Build. Mater.* **2017**, *145*, 253–260. [CrossRef]
31. Kazemian, H.; Naghdali, Z.; Ghaffari Kashani, T.; Farhadi, F. Conversion of high silicon fly ash to Na-P1 zeolite: Alkaline fusion followed by hydrothermal crystallization. *Adv. Powder Technol.* **2010**, *21*, 279–283. [CrossRef]
32. Yijin, L.; Shiqiong, Z.; Jian, Y.; Yingli, G. The effect of fly ash on the fluidity of cement paste, mortar and concrete. In Proceedings of the International Workshop on Sustainable Development of Concrete Technology, Beijing, China, 20–21 May 2004.
33. Sata, V.; Jaturapitakkul, C.; Kiattikomol, K. Influence of pozzolan from various byproduct materials on mechanical properties of high-strength concrete. *Constr. Build. Mater.* **2007**, *21*, 1589–1598. [CrossRef]

MDPI
St. Alban-Anlage 66
4052 Basel
Switzerland
Tel. +41 61 683 77 34
Fax +41 61 302 89 18
www.mdpi.com

Sustainability Editorial Office
E-mail: sustainability@mdpi.com
www.mdpi.com/journal/sustainability

www.ingramcontent.com/pod-product-compliance
Lightning Source LLC
Chambersburg PA
CBHW041347210926
51850CB00013B/3017
9783036535012